JN440086

Cover Image Brother Klaus Field Chapel ©Walter Mair

Published in 2009 by C3 Publishing Co.

72-9 Gonghang Gangseo Seoul 157-812 Korea
Tel. +82 2 2661 2811 Fax. +82 2 2661 2456
E-mail. info@c3-d.com http://www.c3-d.com

Publisher JaeHong Lee 이재홍 Editor Uje Lee 이우재 Managing Editor JaeWon Lee 이재원 Associate Editor JeongSeon Lee 이정선 (Editor in charge)
Editorial Designer MinJung Kim 김민정, HyeKyoung Cheon 천혜경 Color Editor HyunJung Han 한현정 Proofreading EunJung Park 박은정
International Marketing SangMin Lee 이상민 Prepress by pointive 포인티브 Printed by DooKyoung Printech co. Ltd., Seoul 주.두경프린텍

발행처 건축과환경
출판등록 1991년 5월 22일 제13-295호

ISBN 978-89-86780-61-1

Printed in Korea

CREATIVE PERSPECTIVE
in Architecture

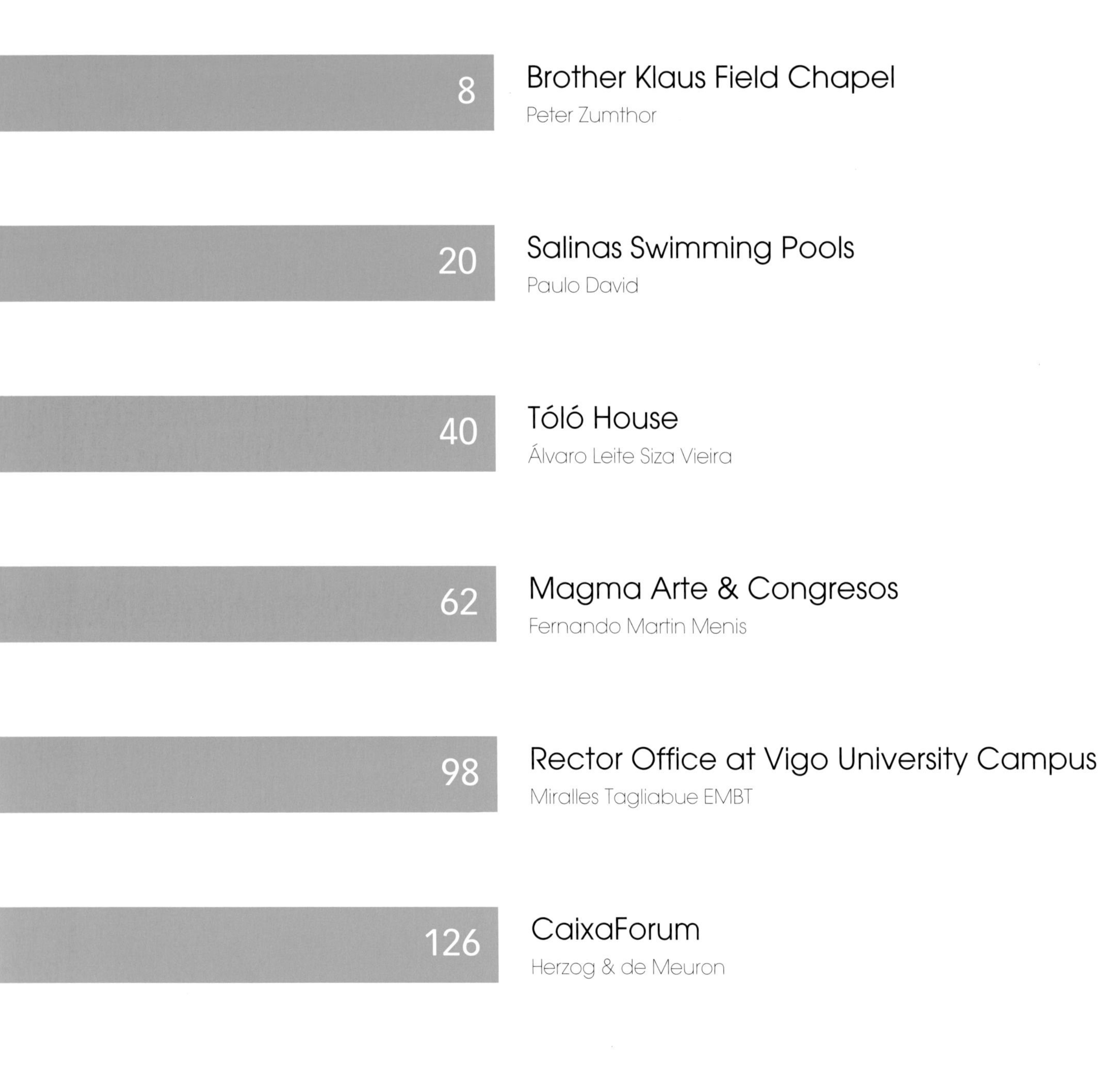

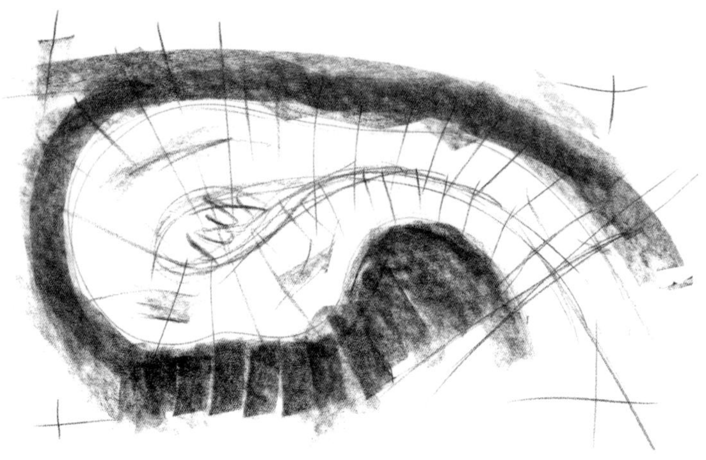

The interior of the chapel was formed by 112 tree logs. The logs have been cut down in the city forest of Bad Muenstereifel and they have been erected by the carpenter Markus Re(c)mann on the field, above a ground plate of concrete, to a tent-like construction. Surrounding the wooden tent, the chapel body made from stamped concrete, was built in layers of 50 cm height. Layer after layer was hoisted up in 24 work days. The concrete is made of gravel from the lake and red-yellow sand from the gravel pit Rhiem in Erp and white cement. The 24 work days have been performed during the time from end of October 2005 to September 2006. For this task, so-called stomping teams were formed by friends and acquaintances of the Scheidtweiler family. They were under the supervision of the builder Alexander Mahlberg and the foreman Hans Joachim Engler. In fall 2006, a special fire was kept burning inside of the wooden tent for three weeks, which let the tree logs dry out and freed them partially from the concrete. Subsequently, the tree logs were dismantled mechanically, a concrete floor with the desired slope was brought in, over which the art caster Miroslav Stransky,

성 클라우스 예배당

예배당의 내부는 112개의 통나무로 이루어져 있다. 이 통나무들은 목수 마쿠스 렉만이 바트 뮌스터라이펠 시의 숲에서 가져온 것으로, 들판 위의 콘크리트 기초 위에 텐트와 같은 모양으로 세워졌다. 이 나무 텐트 주위로 50cm 씩 콘크리트를 쌓아올려 몸체를 만들었다. 한 층을 쌓는 데에는 24일이 걸렸다. 콘크리트는 호수에서 나온 자갈과 자갈채굴장의 주황색 모래, 그리고 흰색 시멘트를 섞어 만들었다. 이 작업은 2005년 10월 말부터 2006년 9월까지 이루어졌다. 이를 위해 샤이트바일러 가는 친지들을 모아 소위 '드림팀'을 구성하여 건설가 알렉산더 말베르그와 현장감독 한스 요아킴 엥글러의 지휘 하에 작업을 진행하였다. 2006년 가을, 나무 텐트 안에 3주 동안 불을 지펴 통나무들을 완전히 태워 건조시키고 조

Peter Zumthor

Brother Klaus Field Chapel

together with his wife Dagmar, casted an approximately 2 cm thick lead floor. Four tons of recycled tin-lead with an extra 6% Antimony were heated gradually in a crucible on the spot and were spread manually over the floor with a ladle.

The locksmith's work, the front door, the sand container for the candles and the little vestry cupboard have been made by the locksmith Willi Muller and by Michael Hamacher, the bench made from one piece of massive lime tree wood, has been made by Markus Re(c)mann, the meditation sign made from brass casting, formed after the picture that Brother Klaus had in his cell as a picture to behold, was produced by Miroslav Stransky. The over 300 stoppers made from crystal glass, which seal the bond openings in the concrete, were mouth-blown in the glass hut Eisch in Frauenau. The head made from bronze, is a work by the Swiss sculptor Hans Josephsohn.

Peter Zumthor is the architect of the chapel, supported by Rainer Weitschies, Michael Hemmi, Frank Furrer, Pavlina Lucas and Rosa Goncalves. Jurg Buchli and Claus Jung were the engineers.

금씩 콘크리트에서 제거하였다. 이후 통나무들은 기계 장비를 사용해 모두 제거되었고, 콘크리트 바닥과 의도했던 경사면만이 남게 되었다. 여기에 주물 예술가인 미로슬라프 슈트란스키와 그의 부인 다그마가 약 2cm 두께의 납 바닥을 주조하였다. 안티몬이 6% 첨가된 4톤에 달하는 재활용 납을 현장의 도가니에 서서히 데우고 쇳바가지를 이용하여 직접 바닥에 부었다. 현관의 모래가 담긴 촛대와 작은 제구 수납장은 자물쇠 제조공 빌리 뮐러와 미카엘 하마커가 제작하였다. 커다란 보리수를 잘라 만든 벤치는 마쿠스 렉만이, 성 클라우스가 방에 두고 보았던 사진을 본따 만든 황동 명상 표식은 미로슬라프 슈트란스키가 제작하였다. 콘크리트의 접합부 구멍을 메우고 있는 300개가 넘는 크리스탈 유리 마개는 프라우에나우의 아이쉬 유리 제조실에서 직접 불어 만든 것이다. 청동 두상은 스위스의 조각가 한스 요셉손이 작업하였다.

Architect Peter Zumthor **Supported** Rainer Weitschies, Michael Hemmi, Frank Furrer, Pavlina Lucas, Rosa Goncalves **Engineer** Jurg Buchli, Claus Jung **Carpenter** Markus Re(c)mann **Builder** Alexander Mahlberg **Foreman** Hans Joachim Engler **Art caster** Miroslav Stransky **Locksmith's work** Willi Muller, Michael Hamacher **Bronze sculptor** Hans Josephsohn **Location** Wachendorf, Germany **Construction** 2005.8 ~ 2006.9 **Photograph** Walter Mair

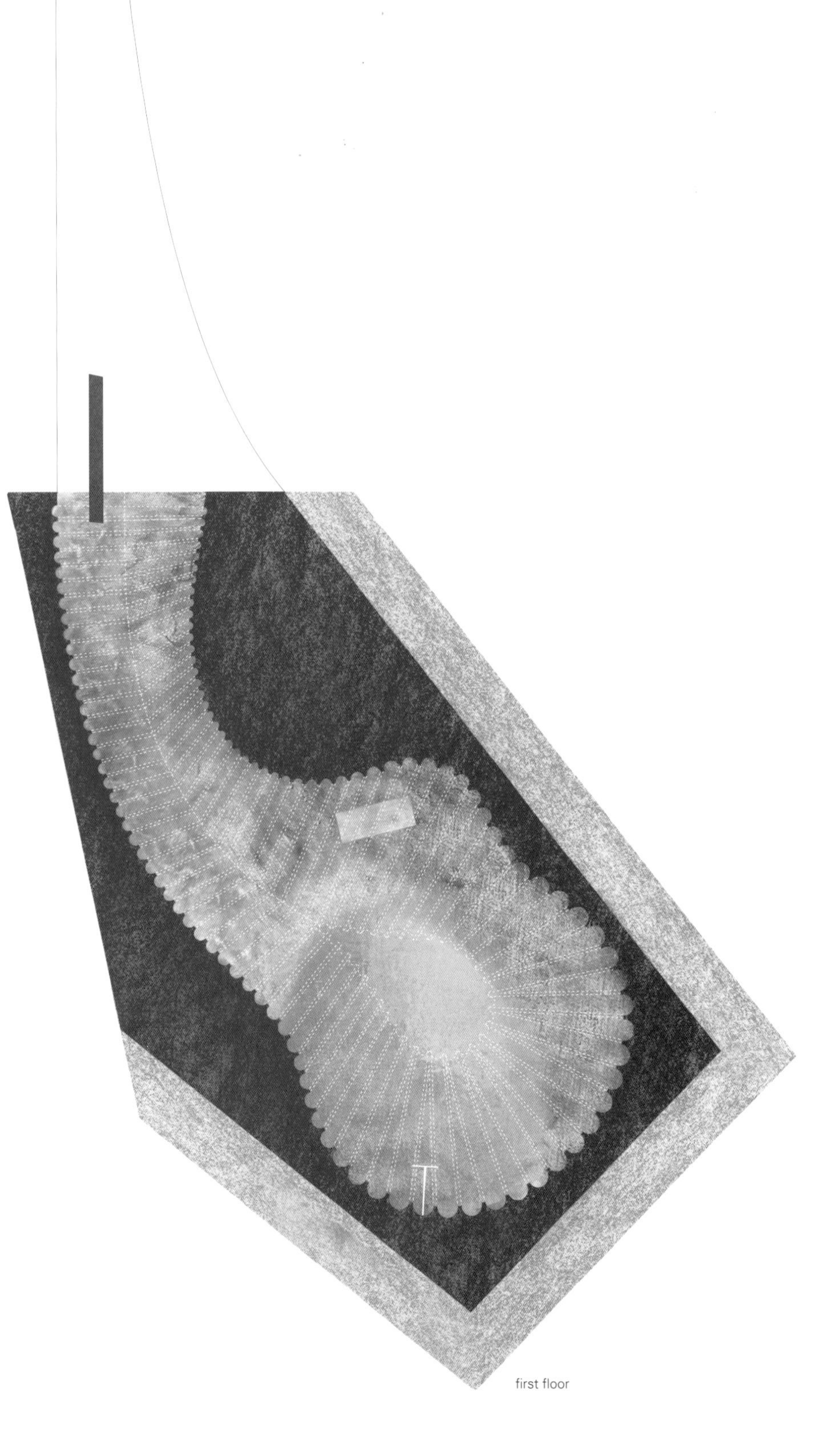

first floor

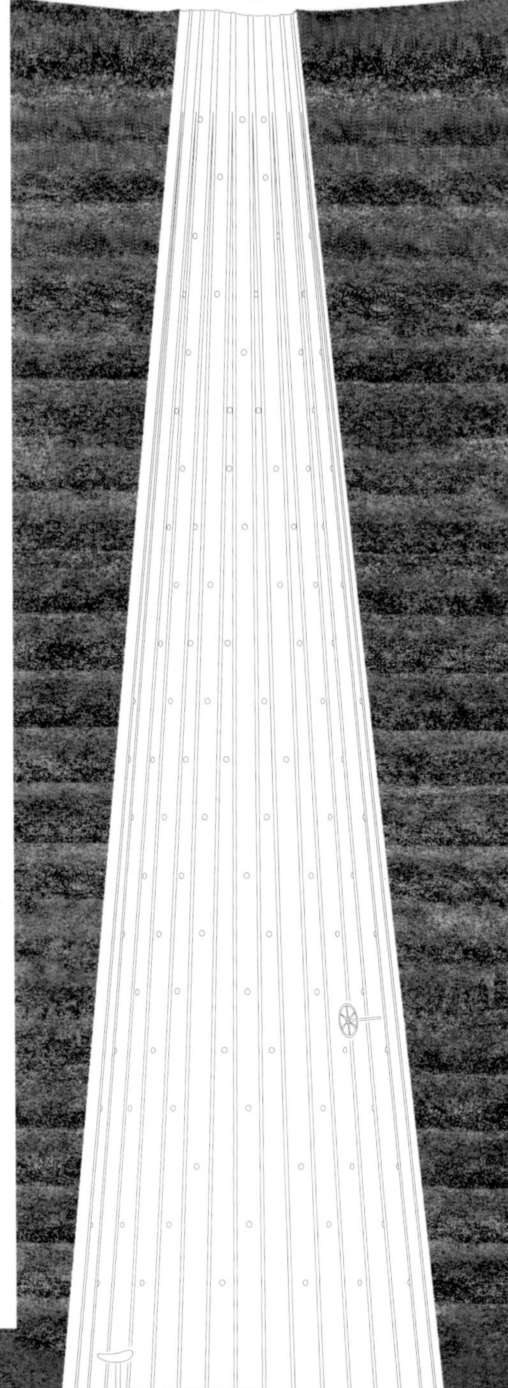

section A - A'

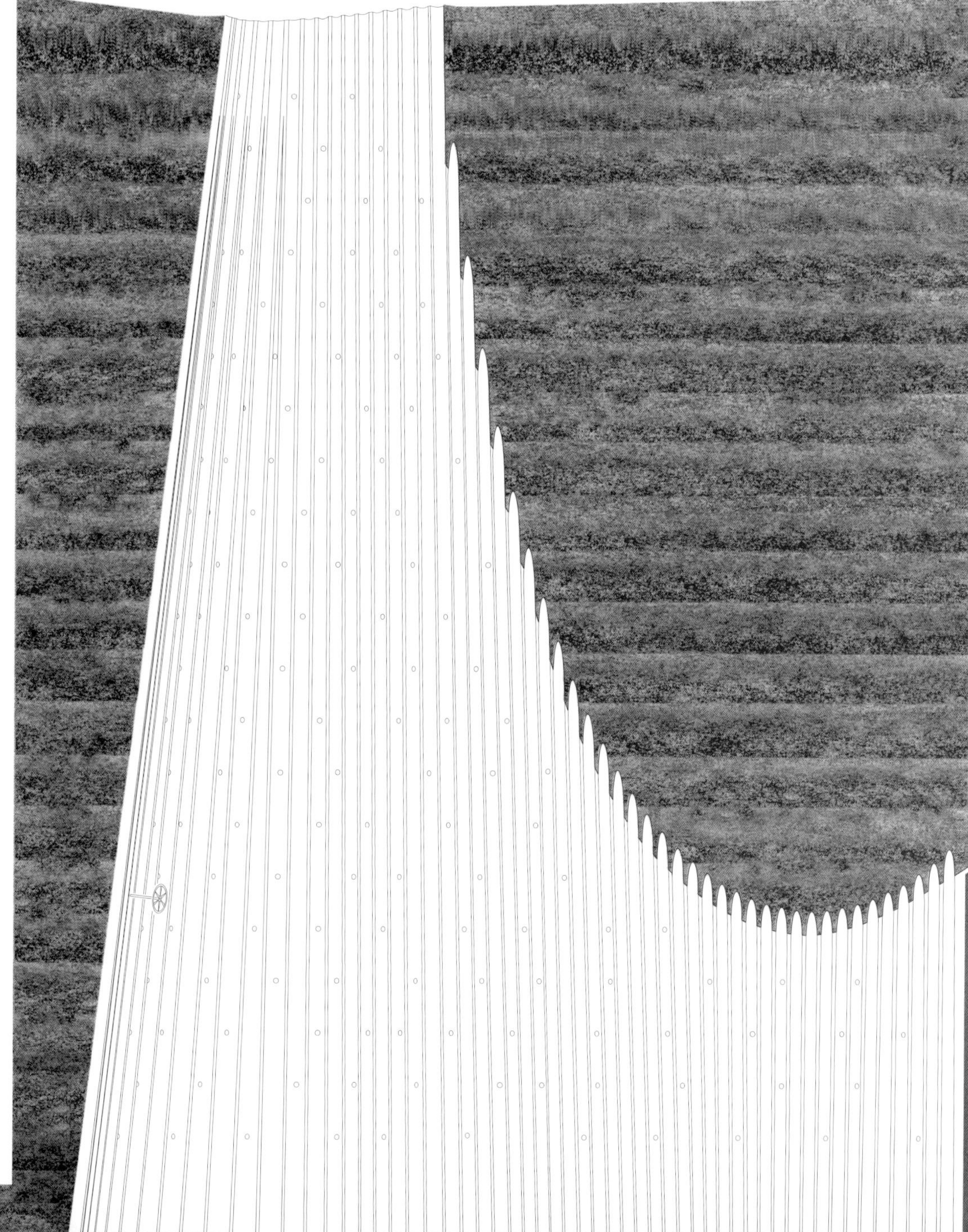

B'
A
A'
B

section B - B'

The place

The Salt Marsh - A construction, the ancient Limekiln.

Configuration - Result of volcanic, black and porous stratification, due to the fast process of lava a rare balance between Form and Place.

Occupation of the Place - An old artisan industry related to fishing industry, dry of 'cat-fish', and extraction of the salt from the rocks gave the name to the place Salinas.

The project

A long wall develops and articulates the peripheral limit of the Salt marsh and gives continuity to a circuit of paths "Caminho da Trincheira" bordering the sea.

A thick wall supports the cliff and shelters the content.

A topographical wall sweetens and limits the surroundings.

A wall in stone contextualizes the place.

A platform in concrete, with precise geometry relates with the sea opposing and confronting the irregularity of the coast evidencing it.

The high quota, the building of the restaurant of intrinsic geometry related to the place, punctuates the horizontality of the thick walls. Creating a view with a panoramic embrasure, that allows a portion of sea within the big room, establishing an intense relation between inside and outside, strengthening the views of the Atlantic.

살리나스 수영장과 해변 산책로 및 레스토랑

장소

염성소택 – 고대 석회가마에 의해 만들어진 지형

지형 – 화산 활동의 결과, 구멍이 많은 검은색 암석 지층이 형성되었고 용암의 빠른 확산으로 형태와 장소가 조화롭지 못하다.

대지의 용도 – 오래된 수산 공장이 하나 있고, 여기저기 말라붙은 메기 잔해들이 있다. 바위에서 소금을 채취하였기 때문에 살리나스라는 지명이 붙었다.

프로젝트

긴 벽이 염성소택의 가장자리를 둘러싸고, 바다와 경계를 이루는 까밍뉴 다 뜨림셰이라 순환로와 이어진다.

두꺼운 벽은 절벽을 지지하고 보호한다.

지형을 따라 만들어진 벽은 주위를 부드럽게 하고 경계를 설정한다.

돌담은 대지의 컨텍스트를 나타낸다.

콘크리트 플랫폼의 기하학적 형태는 마주보는 바다에 부합하며 불규칙한 해안선과 대조된다.

부지에 의해 고유한 형태를 가지는 레스토랑은 두꺼운 수평벽에 마침표를 찍는다. 이곳은 긴 창으로 멋진 조망을 연출하였으며 가장 큰 방에서는 바다를 바라볼 수 있도록 하였다. 그 결과 안과 밖의 관계가 긴밀해지고 대서양으로 탁 트인 전망을 가지게 되었다.

Paulo David

Salinas Swimming Pools

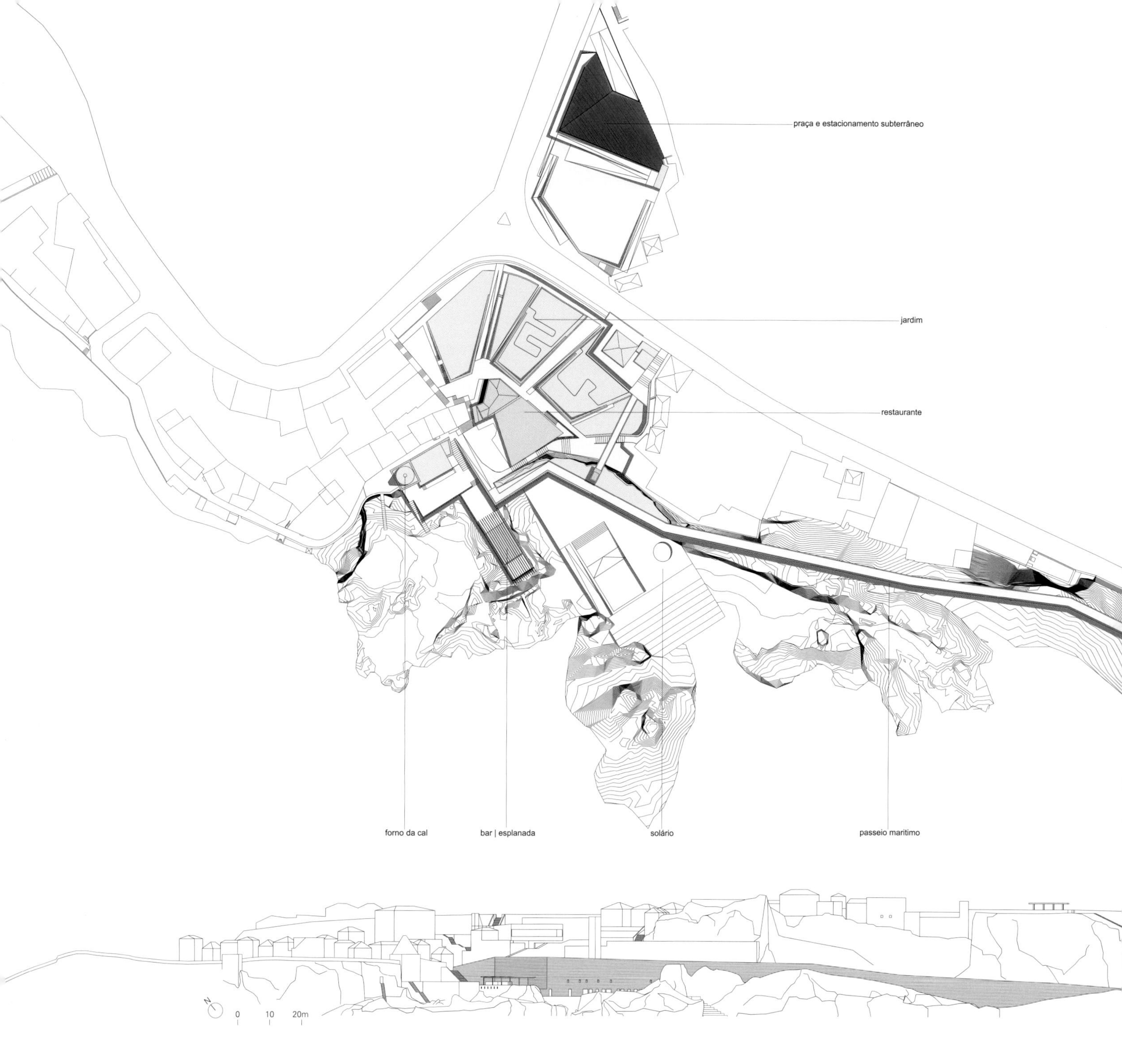

Architect Paulo David **Collaborator** Carlos Aguiar, João Nobrega, Inês Rocha, Dirk Mayer, Patricia Faria, Filipa Tomaz, Rodolfo Reis, Silvia Arriegas, Luis Spranger, Luz Ramalho **Landscape architect** João Gomes da Silva; Global **Structural engineer** Duarte Rui Gouveia; Kplano **Electricity** Ruben Sobral **Climatization** José Galvão Teles **Waters and Sewage** José Jesus **Security** António Matias **Construction company** Consórcio Etermar - Somague **Coordination** ECG Plan - Eng. Rui Canastro **Client** Sociedade Metropolitana de Desenvolvimento, s.a., Vice-Presidência, Governo Regional da Madeira, Portugal **Location** Salinas, Câmara de Lobos, Madeira, Portugal **Project period** 2002~2004 **Photograph** FG + SG Fotografia de Arquitectura

section A - A′
section B - B′
section C - C′
section D - D′
section E - E′
0
10
20m

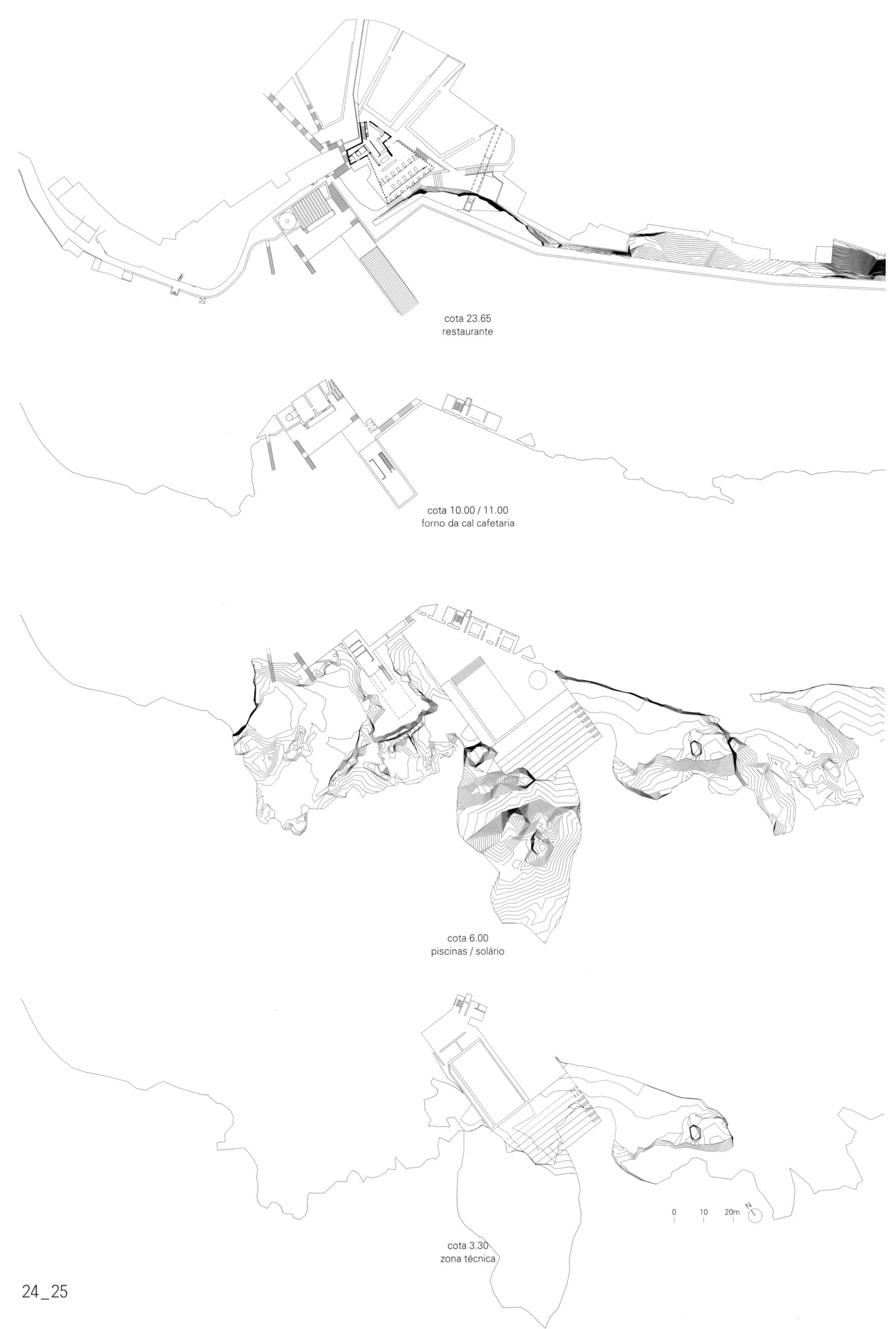

cota 23.65
restaurante
cota 10.00 / 11.00
forno da cal cafetaria
cota 6.00
piscinas / solário
0
10
20m
N
cota 3.30
zona técnica

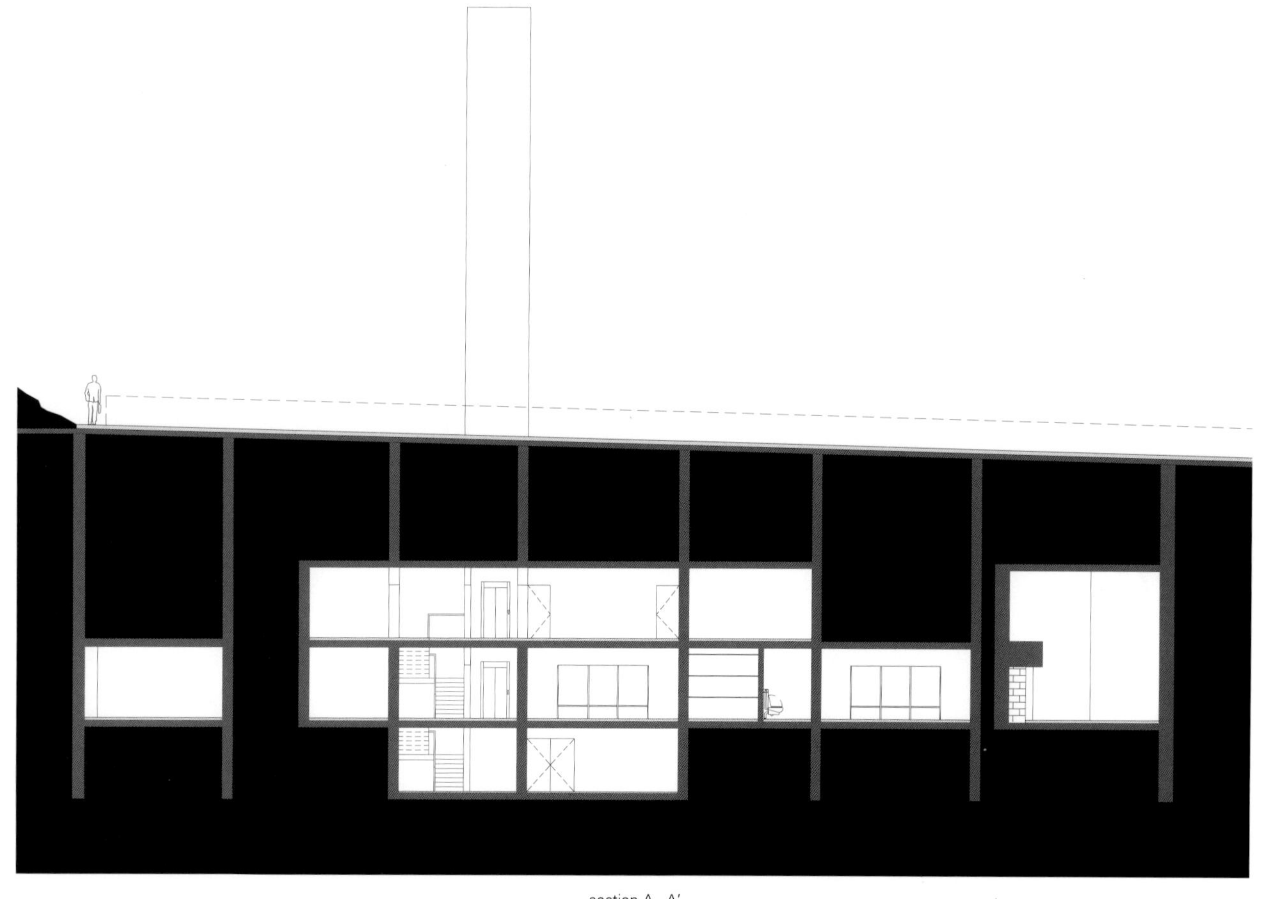

section A - A'

This project is located on a site, with an area of around 1,000m², in Lugar das Carvalhinhas - Alvite, Cerva Parish, Ribeira da Pena Commune, Vila Real District. The project consists of a holiday home with three bedrooms, a social bathroom, a living room, a dinning room, a small kitchen with a support washbasin, pantry, and even a small outdoor swimming pool. The terrain is sharply inclined and has a particular configuration: it is very long and narrow. However, the fact that it faces south allows it to receive optimal solar exposure and to enjoy a particular natural view. The living area does not exceed 180m².

For reasons of functionality and integration, it was opted to organize the main entrance based on the street where car transit was possible. This road leads to the northern higher part of the lot. Nevertheless, it is possible to access the house from a more rustic footpath from the south. Its fragmentation, necessary due to the steep topography, transforms the whole into a composition of small linked and interconnected volumes, creating an unevenness that allows for a more secure and rational use of the lot. In this way the house's various functions are clarified with each elevation corresponding to a single compartment. The roof functions simultaneously as pavement support for the gardens: similar to the traditional threshing floors and patios in the northern regions of the country with hilly terrain.

With a linear position at the center of the lot, an attempt was made to preserve all the preexisting trees, as they maintain a strong presence in the area, as well as to preserve continuity with the immediate surroundings and to ensure its original characteristics. The form resulting from a rigorous, modular geometric abstraction establishes the necessary rotation of certain modules to adapt to the natural morphology of the terrain, respecting the distance from neighboring regulation walls, thus appearing to move naturally and with absolute freedom.

The terrain's instability together with modest economic resources gave rise to a house with triple function: the home itself and its interior space; the organization of outdoor leveled patios corresponding to the roofs of various levels which permit the outdoor use of the garden; and finally, the creation of a pedestrian link between the paths of the upper and lower levels that border this lot. In this way the house itself is a path. Its form organizes a fundamen-

똘루 주택

이 프로젝트는 루가르 다스 까르발링냐스 지역의 약 1천m² 부지에 있는 세 개의 침실, 욕실, 거실, 식당, 싱크대와 작은 부엌, 작은 옥외 수영장 등으로 구성된 주말 주택이다.

지형은 급하게 경사져 있으며 매우 길고 좁다. 그러나 부지가 남향이기 때문에 햇볕이 충분히 들어오고 독특한 자연 경관을 보여준다. 전용면적은 180m²를 넘지 않는다.

집의 기능성과 통합을 위해 주 출입구를 차량 통행이 가능한 부지 북쪽의 높은 부분으로 정하였다. 또한 남쪽의 한적한 통로에서도 집으로 접근할 수 있다. 가파른 지형을 따라 부지를 구획했기 때문에 전체 건물은 서로 연결된 볼륨의 집합체이며, 그 결과 건물이 불규칙하게 구성되어 부지를 더욱 안전하고 합리적으로 사용될 수 있게 되었다. 이 때문에 주택의 다양한 기능은 고유의 입면을 가지며 지붕은 정원으로도 사용할 수 있다. 이러한 구조는 구릉지가 많은 포르투갈 북부지역에서 흔히 볼 수 있는 전통적인 탈곡장과 유사하다.

기존에 있던 나무들은 인접 지역과의 연속성과 부지 고유의 특징을 유지할 수 있도록 부지의 중앙선을 따라 모두 남겨두었다. 엄격한 기하학적 모듈에서 나온 형태를 필요한 지점에서 회전시켜, 주위의 규칙적인 벽들과 거리를 유지하면서 지형을 따르도록 하였다. 그 결과 이러한 모듈은 자연스럽고 자유롭게 움직이는 것처럼 보인다.

불안정한 지형과 한정된 예산 속에서 세 가지 기능을 가진 주택이 만들어졌다. 한 가지는 주택 자체와 내부 공간이며, 다른 기능은 다양한 높이를 가진 지붕 위의 옥외 중정이고, 나머지는 부지의 경계인 위와 아래 통로 사이의 보행로이다. 이

Álvaro Leite Siza Vieira

Tóló House

tal outdoor route. The exterior stairs that create a link between the patios mirror the interior stairs that have the same function of linking the designed compartments that are also developed through levels. In this way, the exterior stairs correspond to the interior ceilings.

Although this project presents an extreme unit in its language, from the four quadrants, its image varies completely. It expresses its entire volumetry to the southeast, and in a different way to the northeast. It is radically contextualized in the northwest and southwest quadrants. It is neither horizontal nor vertical architecture. Its location on the lot results in a slanting architecture.

The necessity of partially burying the house comes from lack of economic means, and in this way creates positive thermal behavior and security. The outdoor patios solidly adapt, establishing a direct link to the garden, and the house is naturally unified with the terrain. The choice of exposed concrete creates an idea like that of massive stones appearing naturally on the site. In this way, expressiveness is extracted from a continuous structure of reinforced concrete, the most efficient on the lot with these characteristics, and once again making optimum use of the modest economic resources available.

Paving stones of reinforced concrete will also be constructed, creating an air space between these and the lot, which is fundamental for guaranteeing decent waterproofing and improved thermal behavior in the terrain with such an elevated water table. The reinforced concrete walls will contain an exterior finishing with special formwork, its stereotomy having been designed. PVC canvases will waterproof foundations, walls, and roofs.

The roofing will be covered with thermal insulation and prefabricated, anti-slip tiles. Some patios will be filled with humus for a grass plane. The few non-resistant walls will be built in cement blocks filled with sand. These will be plastered, coated and painted white in the interiors.

The interior floors, doors, and baseboards are in wood, except in the water areas. The exterior doors and windows are in metal, which are double-pained for improved thermal and acoustic insulation.

로써 주택 자체가 하나의 통로이자 옥외 통로의 기본 요소가 된다. 중정을 연결하는 외부 계단은 다양한 높이의 내부 공간을 연결하는 내부 계단을 반영하며 외부 계단은 내부의 천장에 해당한다.

모든 방향에서 극단적인 건축 구성을 보여주는 이 프로젝트는 매우 다양한 이미지를 가진다. 동남쪽으로는 전체 볼륨을 드러내며, 동북쪽으로는 다른 방법으로 볼륨을 드러낸다. 북서와 남서쪽의 컨텍스트는 극단적이다. 프로젝트는 부지로 인해 수평적인 건축도 수직적인 건축도 아닌, 경사진 건물이 되었다.

경제적인 이유로 주택을 부분적으로 지하에 묻었기 때문에 보온과 안전성 면에서 효율적이다. 옥외 중정은 주변 정원과 바로 이어져, 주택은 자연스럽게 지형과 통합되었다. 노출 콘크리트로 지어진 집은 부지 위에 자연스럽게 모습을 드러낸 바위 같은 인상을 준다. 이처럼 철근 콘크리트 구조의 효과적인 표현으로 이 부지에 가장 효율적이면서 부족한 경제적 자원을 적절히 사용한 사례가 되었다.

철근 콘크리트의 보행로와 주택 사이의 공간은 해발이 높은 지형에서 방수와 보온에 효과적이다. 콘크리트 벽은 특수한 거푸집 공사를 통해 외부 마감이 되도록 절단 방식을 미리 설계하였다. PVC 캔버스는 연못, 벽, 지붕 등을 방수한다.

지붕은 단열재와 공장에서 제작된 미끄럼 방지 타일로 덮을 예정이며 일부 중정은 녹지 조성을 위해 부식토로 채워진다. 모래로 채운 시멘트 블록에 몇 개의 비내력벽을 세우고 회반죽으로 코팅한 후 흰 페인트를 칠할 예정이다. 수공간을 제외한 내부 바닥, 문, 기초 판자는 모두 목재로 마감하였다. 외부 문과 창호는 금속을 사용하였으며, 보온과 방음을 위해 모두 이중으로 하였다.

Architect Álvaro Leite Siza Vieira **Landscape architect** Álvaro Leite Siza Vieira **Stability** Gabinete de Organizacao e Projectos, Lda. **Water and Sewers** Gabinete de Organizacao e Projectos, Lda. **Electrical** Gabinete de Organizacao e Projectos, Lda. **Mechanical equipments** Gabinete de Organizacao e Projectos, Lda. **Acoustics** Gabinete de Organizacao e Projectos, Lda. **Constructor** Óscar Gouveia **Client** Luís Marinho Leite Barbosa da Silva (Tóló) **Location** Lugar das Carvalhinhas, Portugal **Project period** 1999~2005 **Construction area** 180 m^2 **Cost** 150,000 Euros **Photograph** FG + SG Fotografia de Arquitectura

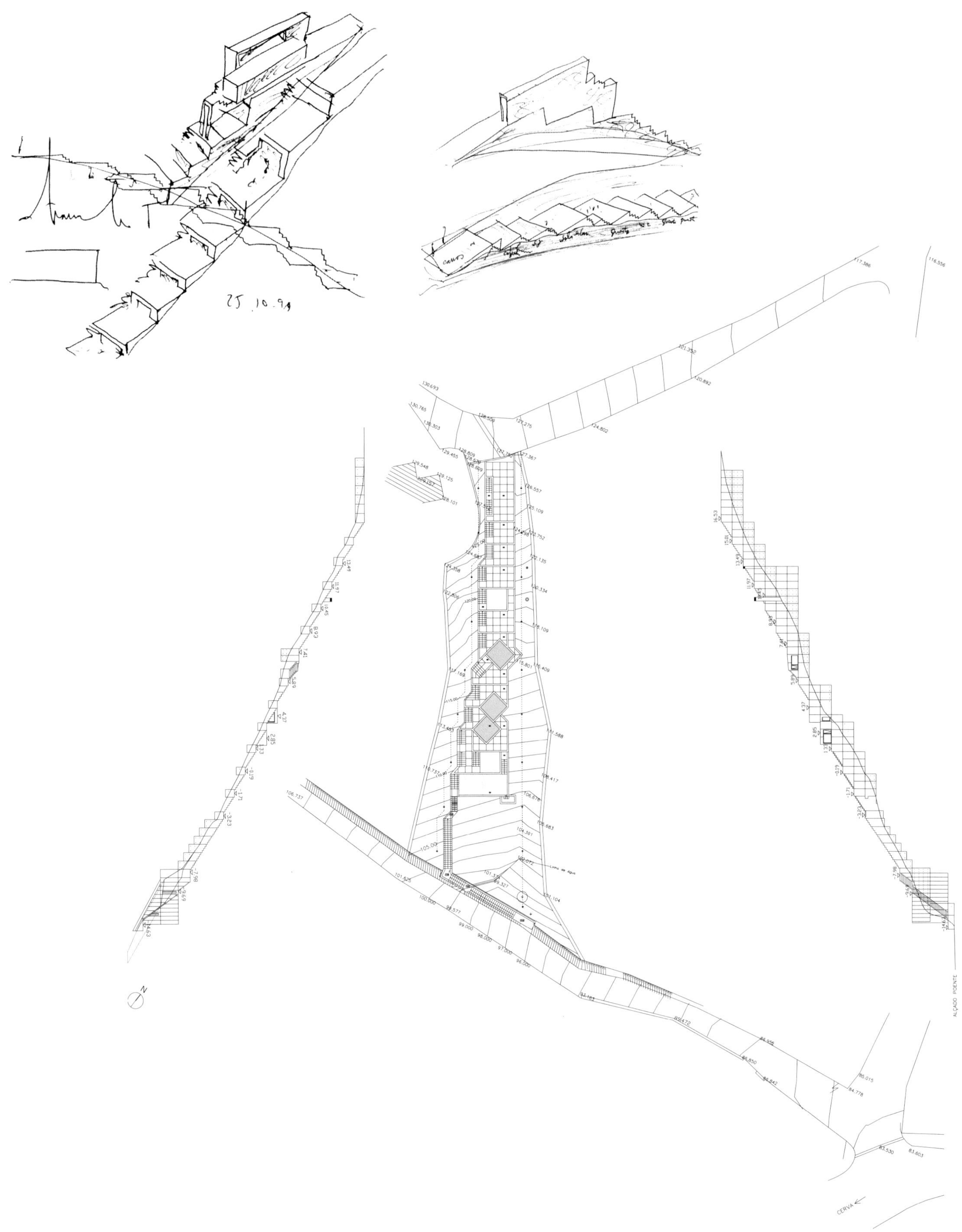

ALÇADO POENTE
CERVA

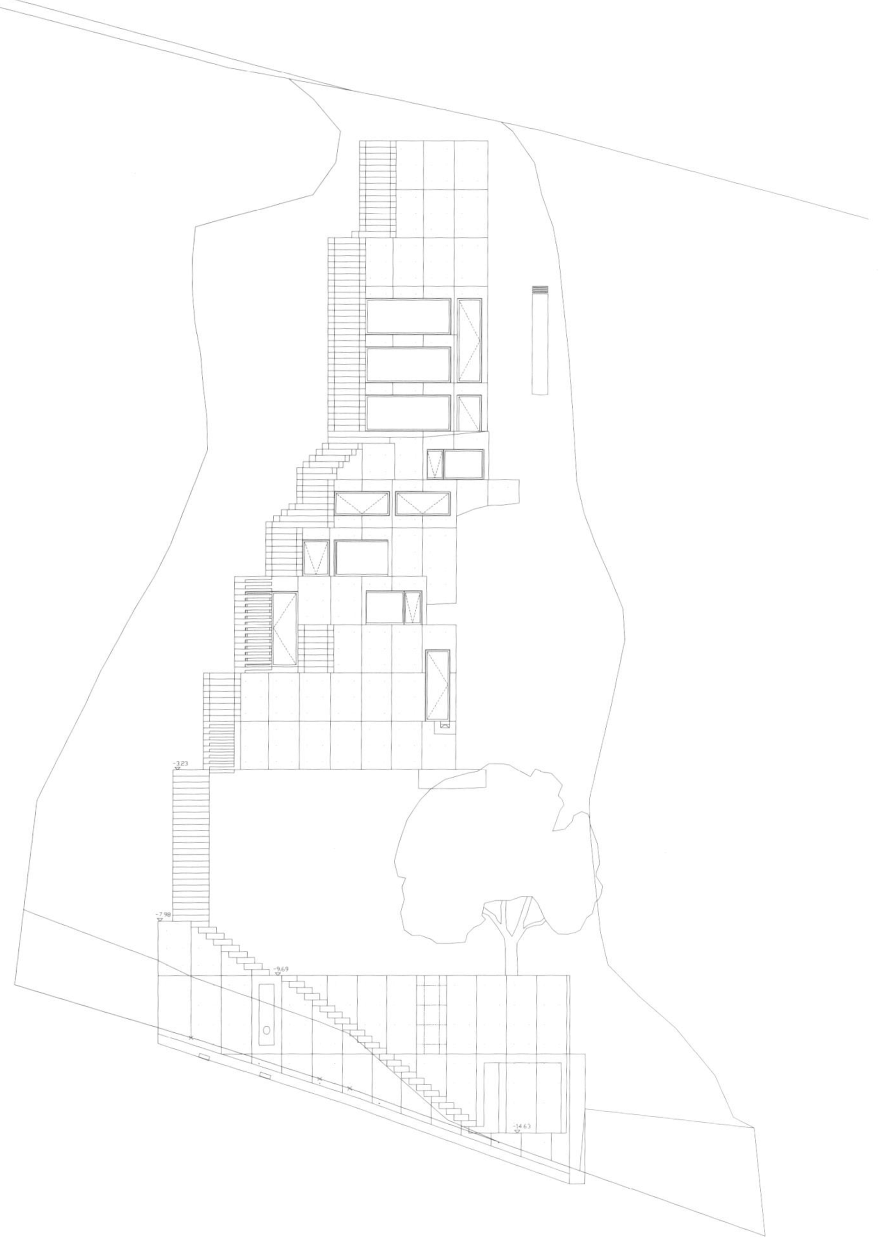

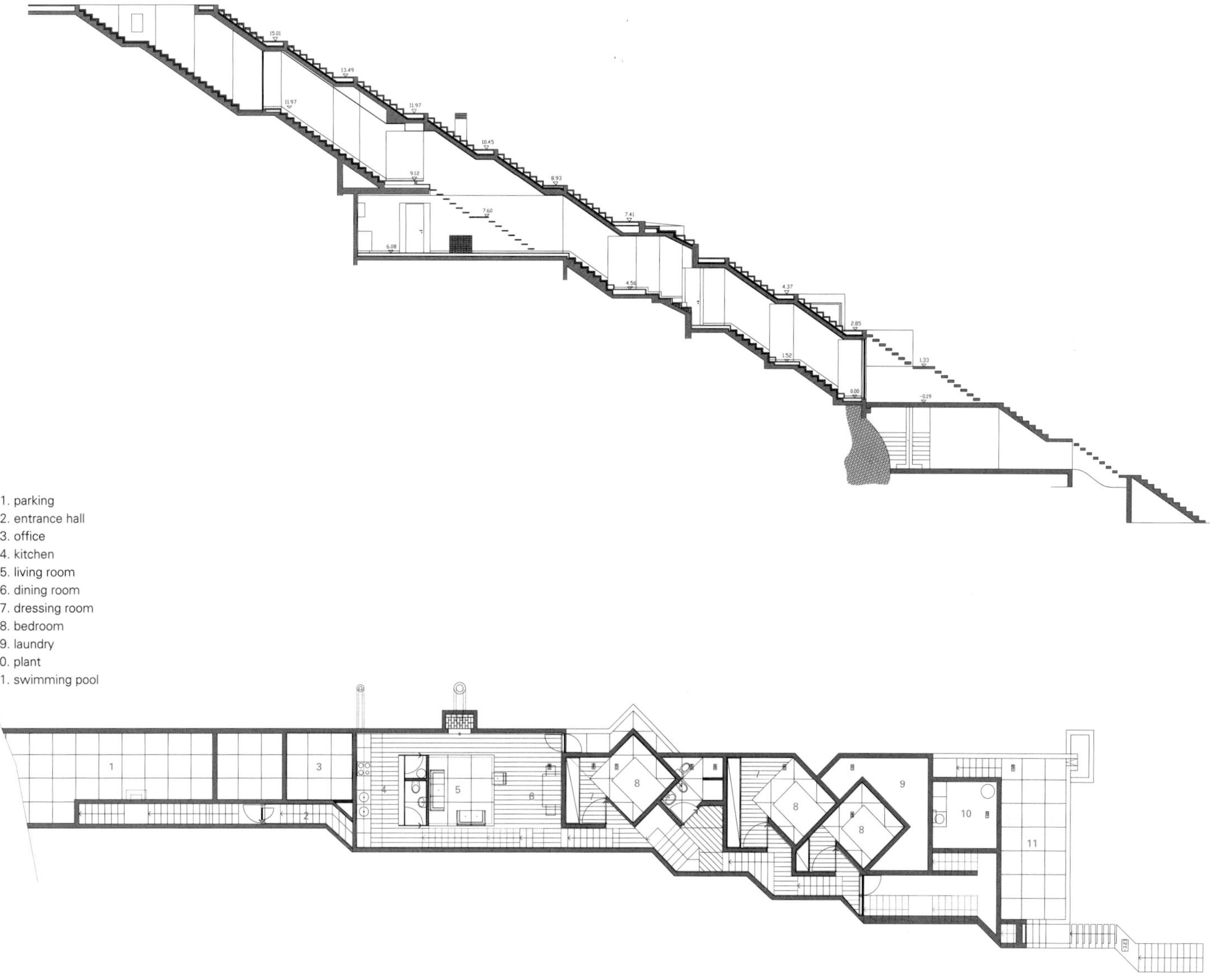

1. parking
2. entrance hall
3. office
4. kitchen
5. living room
6. dining room
7. dressing room
8. bedroom
9. laundry
10. plant
11. swimming pool

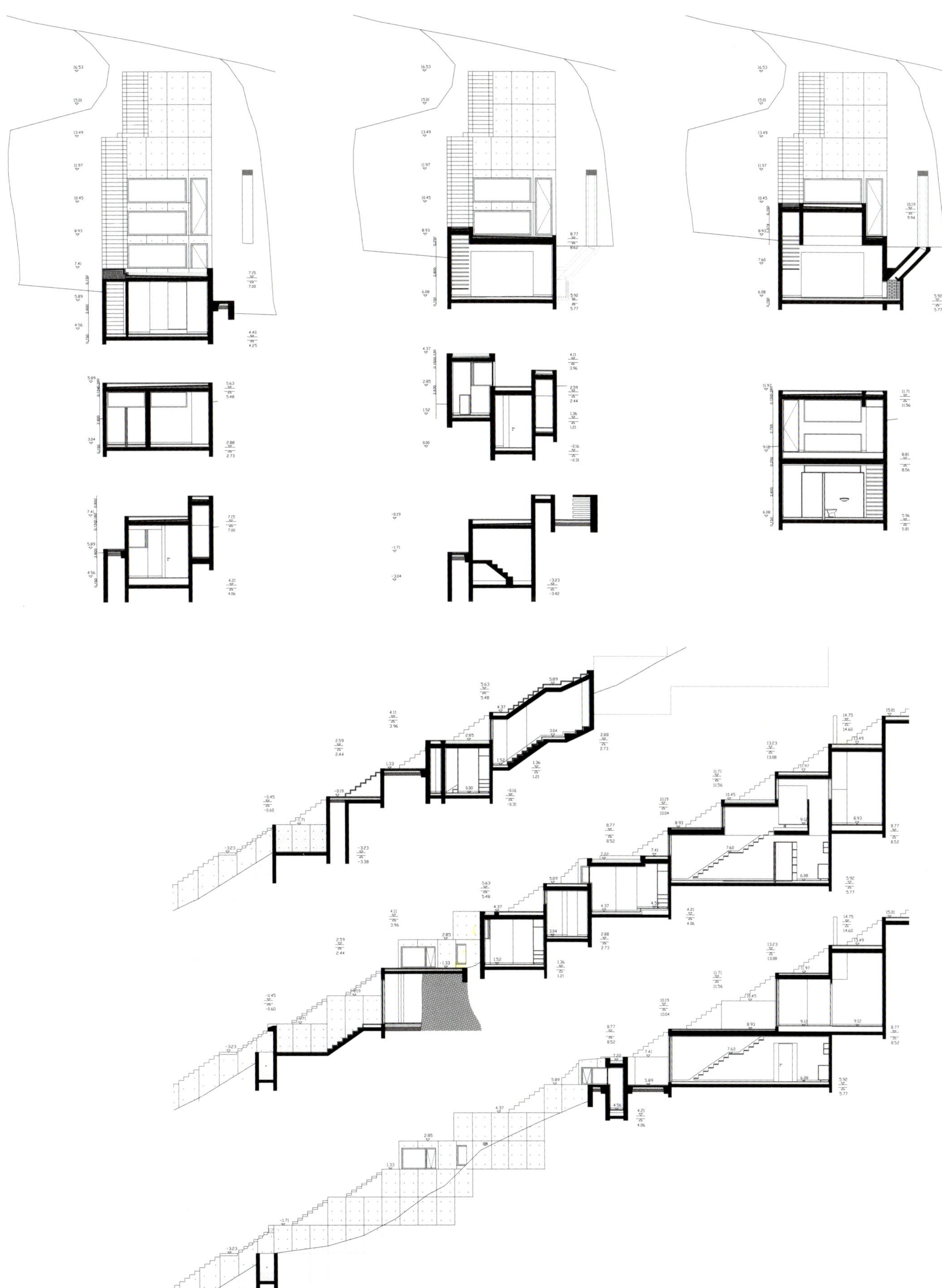

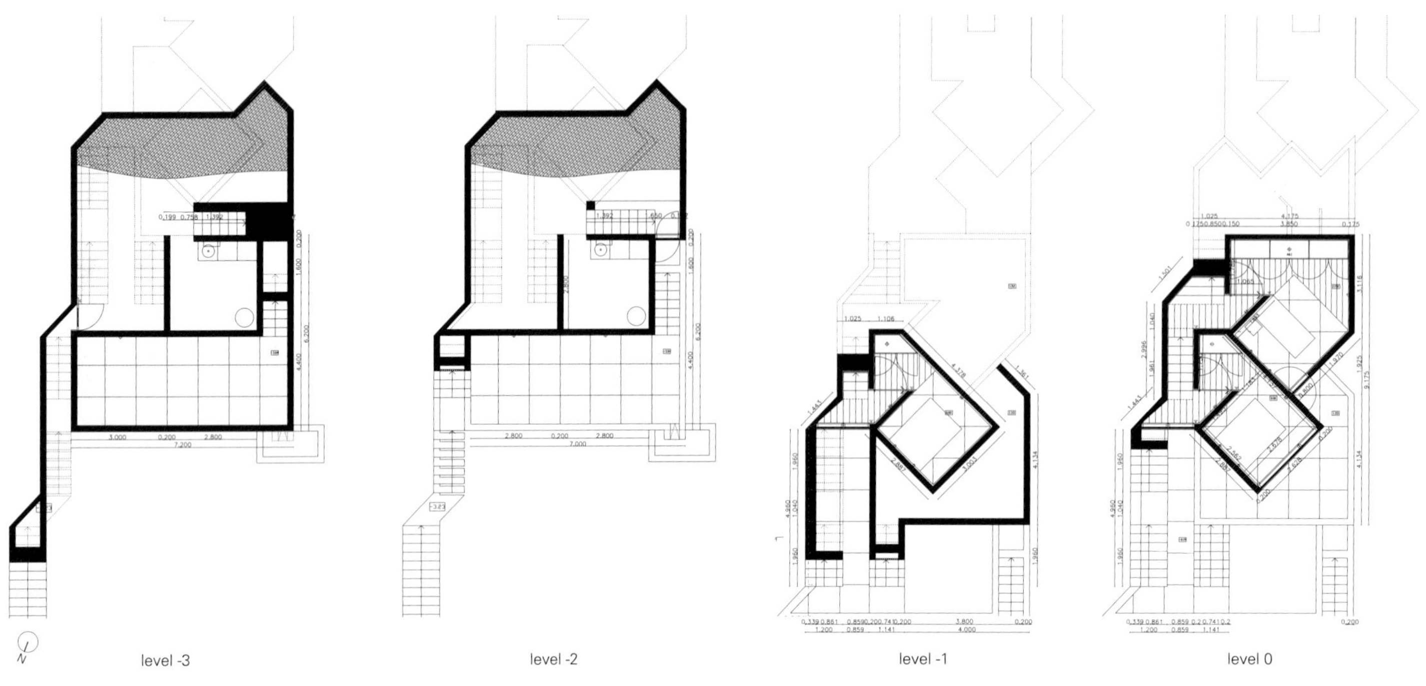
level -3
level -2
level -1
level 0

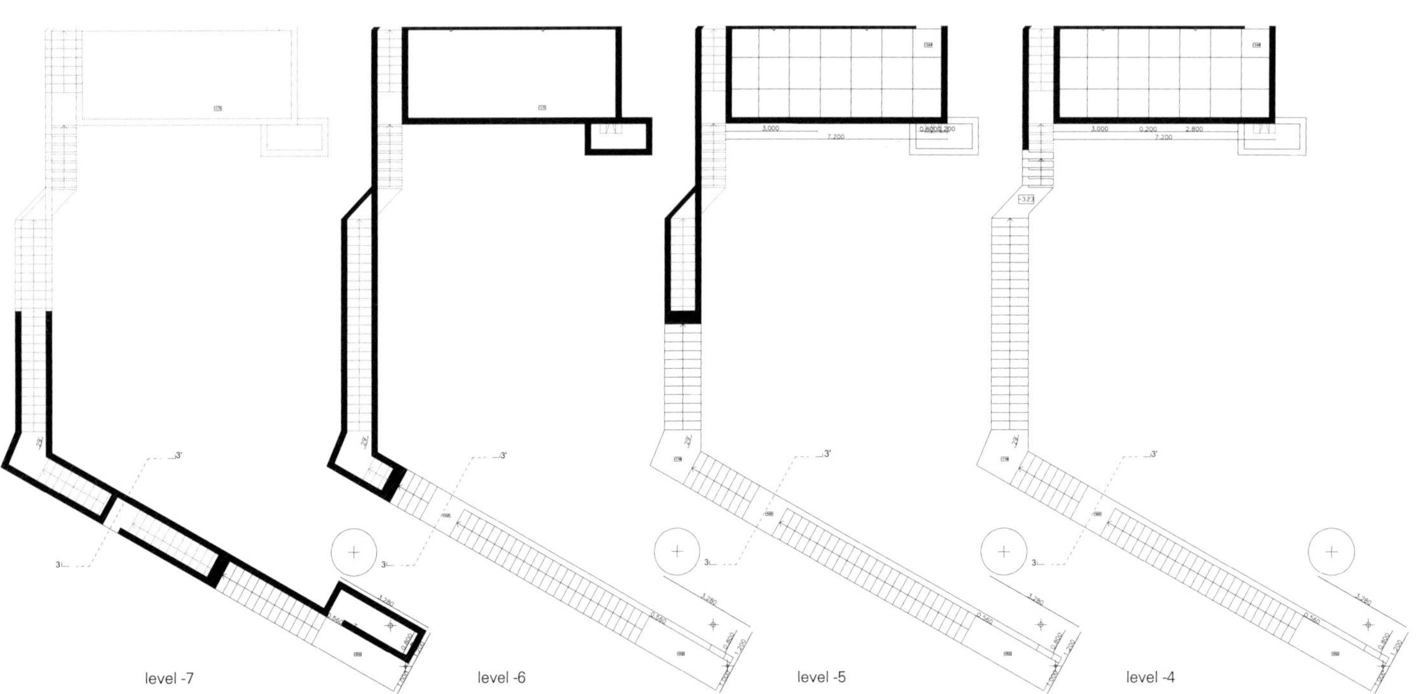
level -7
level -6
level -5
level -4

level 6 level 7 level 8 level 9 level 10

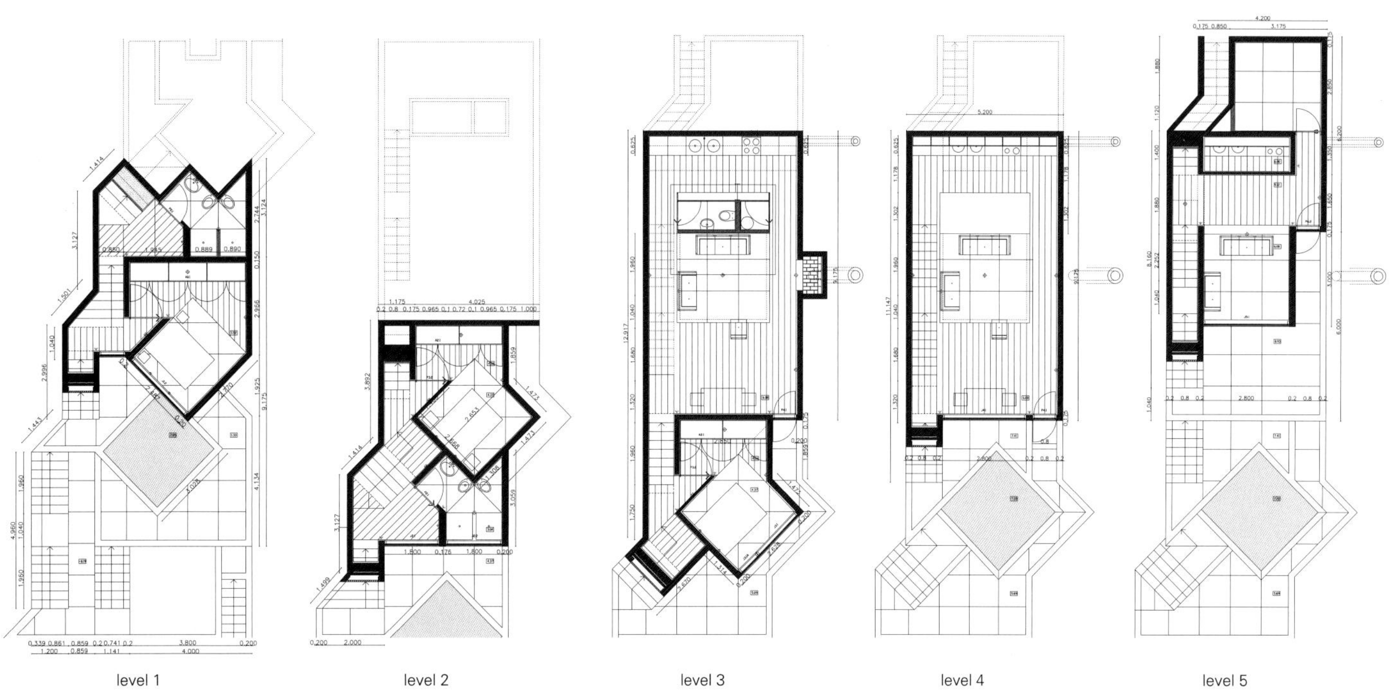

level 1 level 2 level 3 level 4 level 5

Set in an exceptional location close to the airport and highway, and placed in the biggest tourist spot in the island, surrounded by numerous hotel complexes, the MAGMA Arte & Congresos, the Convention Centre in Tenerife, appears as a reference point. The semi-desert landscape around it and the presence of the sea are the starting points of the concept of the building. Though it is not situated directly on the water, it does have a strong relation with the ocean, an imposing presence that frames the building with a constant view of La Gomera Island. The building takes advantage of this relationship, being set on the hill at the perfect height to show off its profile against the sea.
From the distance, the building asserts itself arrogantly through its constructive forms, expressiveness, and strength, while close up it dissolves into its surroundings, blending with the environment. Thirteen geometrically shaped blocks emerge from the base; these house all programmatic functions: offices, restrooms, cafeteria etc These pieces arise, creating a fault line that produces the flow of the roof, imagined as a liquid in motion, outlining the space in

마그마 예술센터 및 회의장

마그마 예술센터 및 회의장은 공항과 고속도로 가까이 있으며 섬의 가장 큰 관광지 내에 수많은 호텔들로 둘러싸여 있다. 떼네리페에 있는 이 컨벤션 센터는 하나의 랜드마크이다.
반 사막의 주변 랜드스케이프와 바다는 건물 설계의 시작점이다. 비록 이 건물은 직접 물과 맞닿아 있지는 않지만, 탁 트인 바다를 배경으로 하고 있어 라 고메라 섬을 건물에 지속적으로 담아내며 긴밀한 관계를 맺는다. 이러한 이점을 살려 건물은 언덕 위에서 바다를 가장 잘 볼 수 있는 높이에 자리잡았다.
멀리서 보면 건물은 그 구성주의적 조형, 표현성 및 힘찬 제스처를 통해 자신을 뽐내고 있는 것처럼 보이며, 가까이에서 보면 주변과 융화되어 환경의 일부처럼 보인다. 땅으로부터 솟아오른 듯한 13개의 기하학적인 블록은 사무실, 화장실, 카

Fernando Martin Menis

Magma Arte & Congresos

every direction. The undulating surface becomes fractured, creating cracks of light and ventilation, splitting and multiplying everywhere with a sensation of lightness. The in-between spaces are the conference halls, which can be used for different activities with varying degrees of compartmentalization.

The main conference hall, placed in the ground floor, has 2,354m² surface plus a stage measuring 300m² that can hold around 2,500 people. It can be divided in up into nine small conference rooms that can seat 300 people each. The first floor hall has 1,865m² of surface and can be also divided into smaller rooms that accommodate between 20 and 200 people. This transformation is achieved by soundproofed panels set in wardrobe - rocks. Technical requirements have determined, which strives for certain acoustic results. The flowing and undulating role of the roof inside responds to technical necessities which were determined during the construction process, with the help of collaborators, among them the director of the Symphonic Orchestra of Tenerife.

페 등 다양한 프로그램을 수용하고 있다. 이 블록이 솟아오르면서 유동적인 액체와 같은 지붕의 단층선을 형성하며 모든 방향에서 공간의 윤곽을 만들어낸다. 이 물결치는 지붕의 표면이 갈라지면서 자연광과 통풍을 위한 틈을 형성하며 빛의 느낌을 분할, 증폭한다.

중앙의 회의실은 칸막이의 각도를 다양하게 변경하여 여러 가지 용도로 쓰일 수 있다. 약 2,500명의 사람들을 수용할 수 있는 1층의 주 회의실은 300m²의 무대를 포함하여 전체 면적이 2,354m²이며 각각 300명을 수용할 수 있는 9개의 소회의실로 분할될 수 있다. 1,865m²의 2층 홀 역시 라커룸이 내장된 방음 패널로 20명에서 200명까지 수용 가능한 소회의실로 나눠질 수 있다. 이곳에는 음향효과를 위해 다양한 기술들이 적용되었다. 물결치듯 흐르는 지붕의 안쪽 면은 시공 과정에서 떼네리페 심포니 오케스트라 단장을 비롯한 여러 사람들의 자문을 얻어 결정되었다.

Architect Fernando Martin Menis, Felipe Artengo Rufino, Jose Maria Rodriguez Pastrana **Assistant architect** Esther Ceballos (2000-2005), Andreas Weihnacht (1997-2000), Ana Salinas Mata (urbanization) **Engineer** Rafael Hernandez, Andrés Pedreno **Technical consulting** Victor Martinez Segovia (structure), Juan José Gallardo (urban environment structure), Milian Associats SL (electricity project), Audioscan SL (acustic planning), Carlos Belda (lighting planning), Juan Gopar (lamps design) **Constructor** NECSO, PPL, UTE CONGRESS **Client** Canarias Congress Bureau Tenerife Sur **Location** Tenerife, Canary Island, Spain **Bldg. area** 20,434 m² **Project period** 1998~2005 **Completion** 2005 **Cost** 25M Euros **Photograph** Torben Eskerod

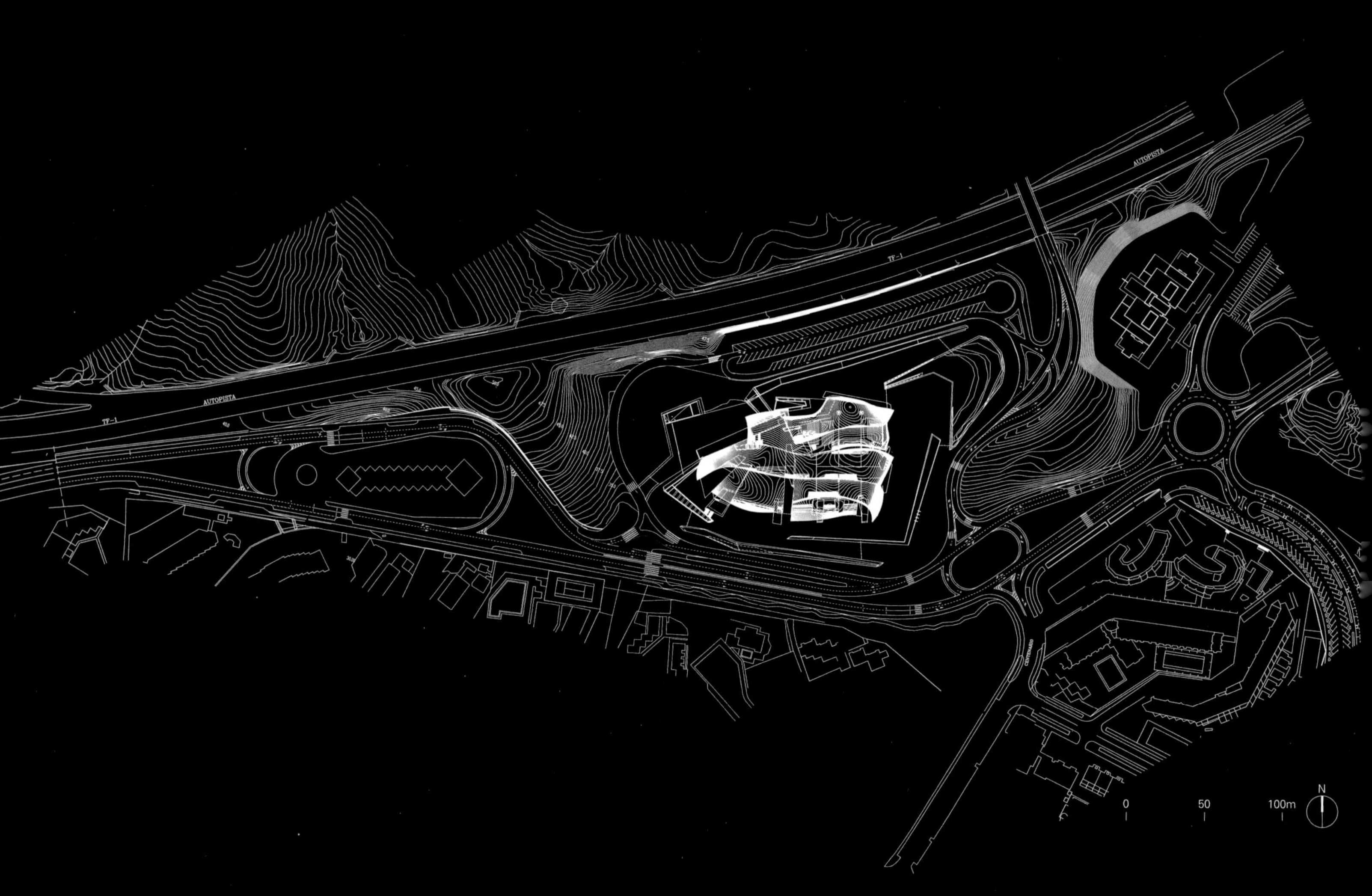
AUTOPISTA
TF-1
AUTOPISTA
TF-1
0
50
100m
N

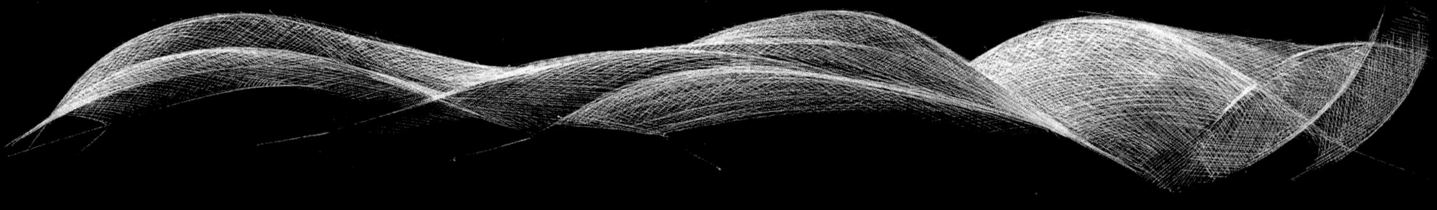

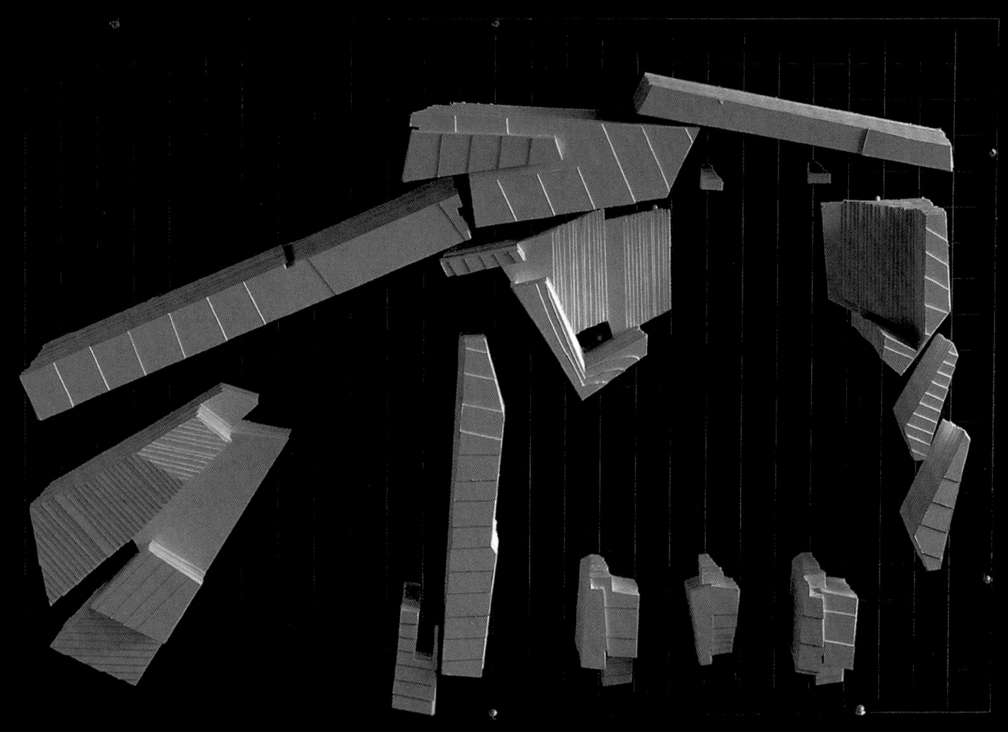

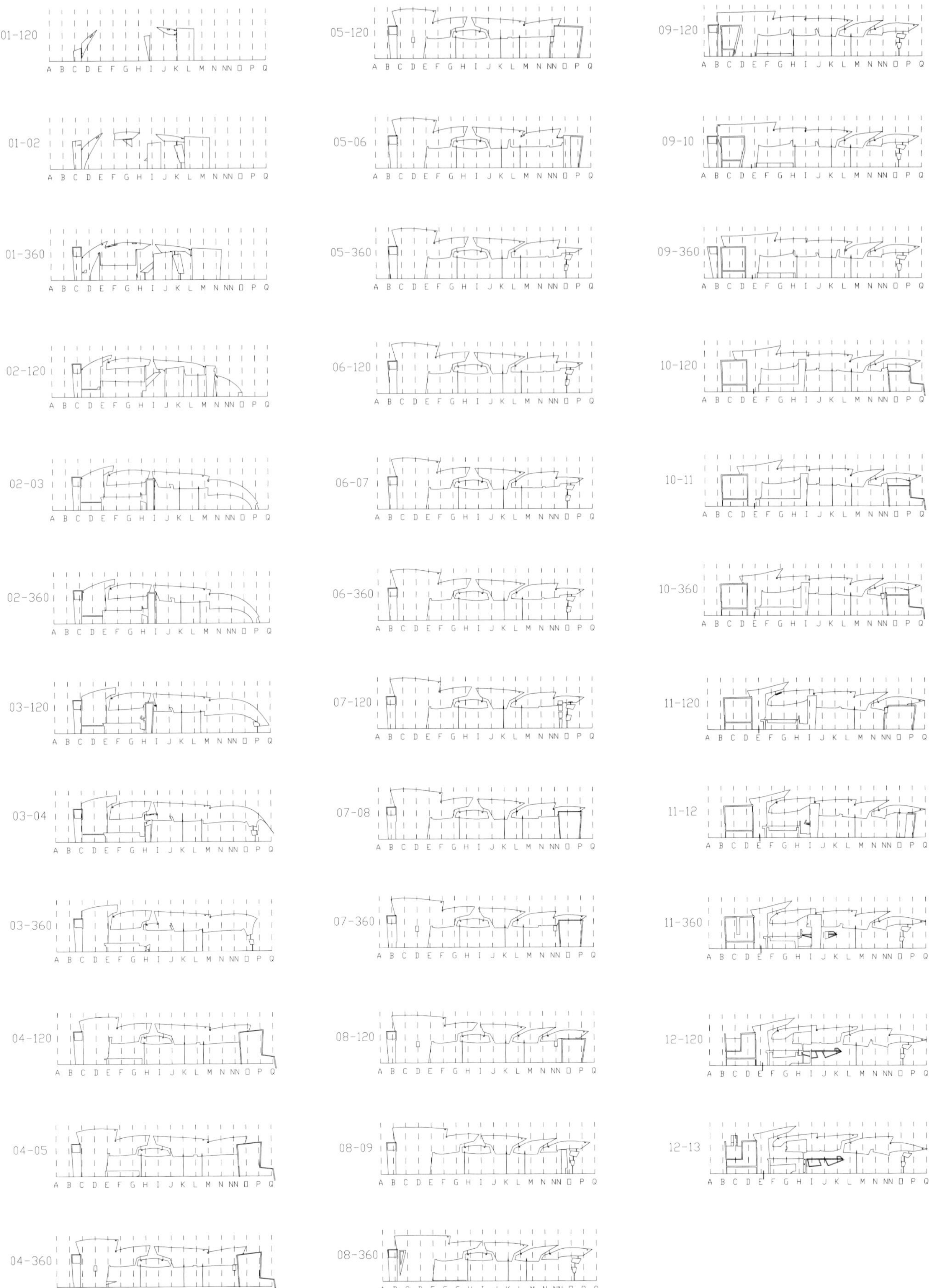

01-120
01-02
01-360
02-120
02-03
02-360
03-120
03-04
03-360
04-120
04-05
04-360
05-120
05-06
05-360
06-120
06-07
06-360
07-120
07-08
07-360
08-120
08-09
08-360
09-120
09-10
09-360
10-120
10-11
10-360
11-120
11-12
11-360
12-120
12-13
A B C D E F G H I J K L M N NN O P Q

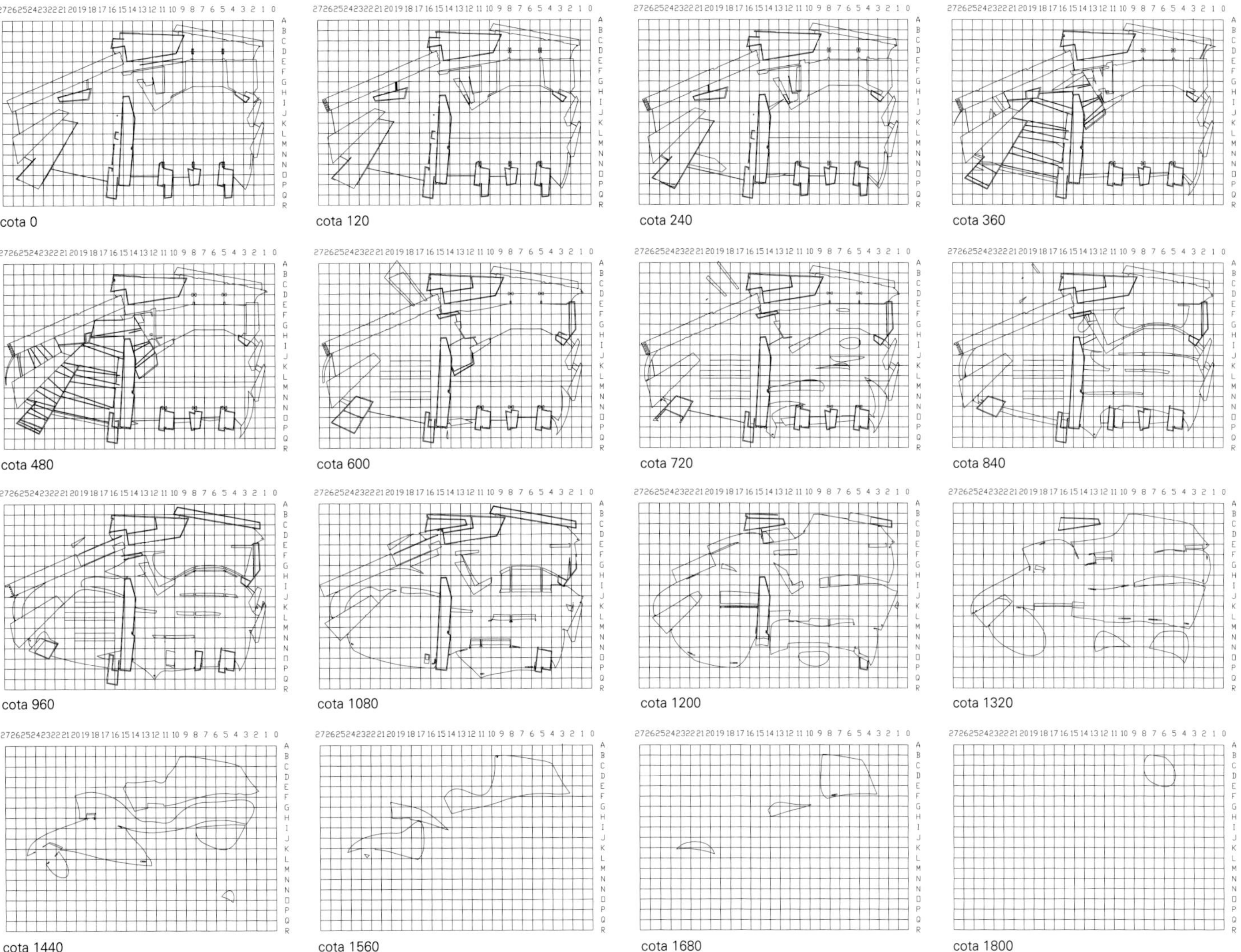
cota 0
cota 120
cota 240
cota 360
cota 480
cota 600
cota 720
cota 840
cota 960
cota 1080
cota 1200
cota 1320
cota 1440
cota 1560
cota 1680
cota 1800

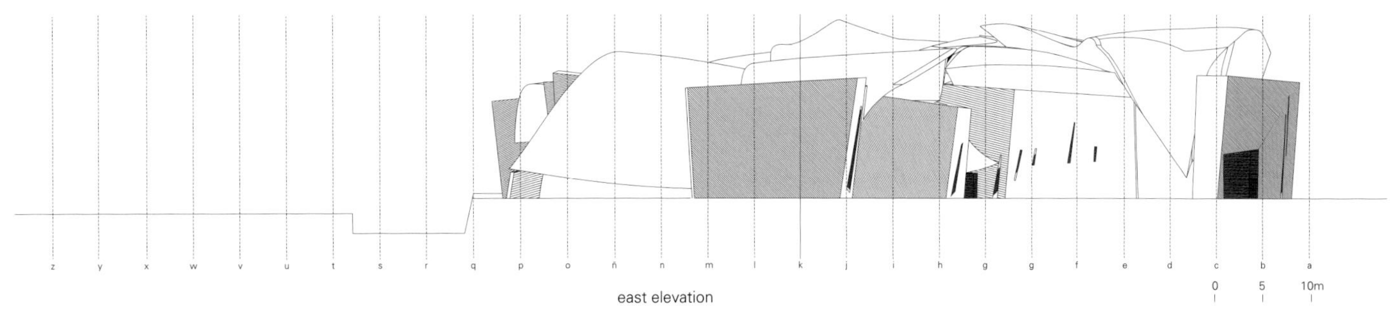

east elevation

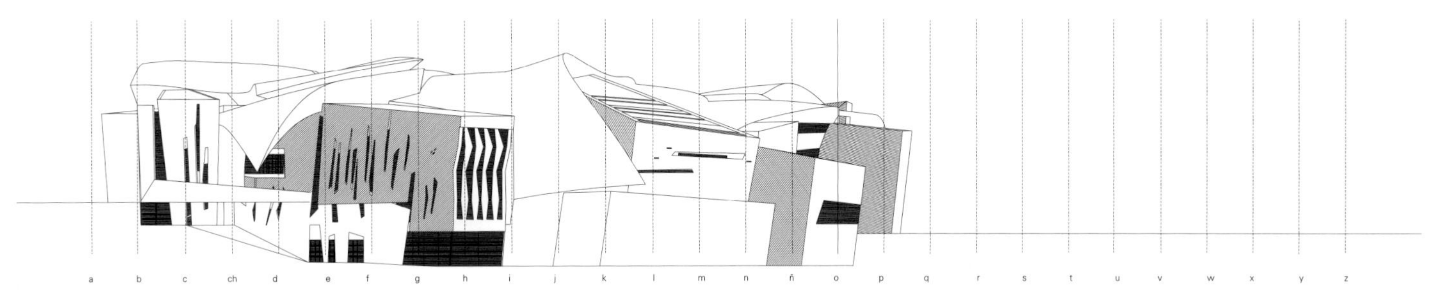

west elevation

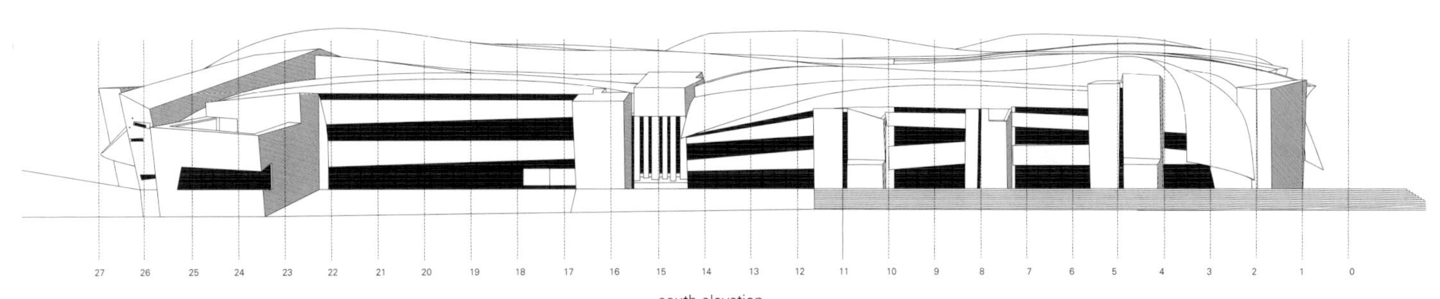

south elevation

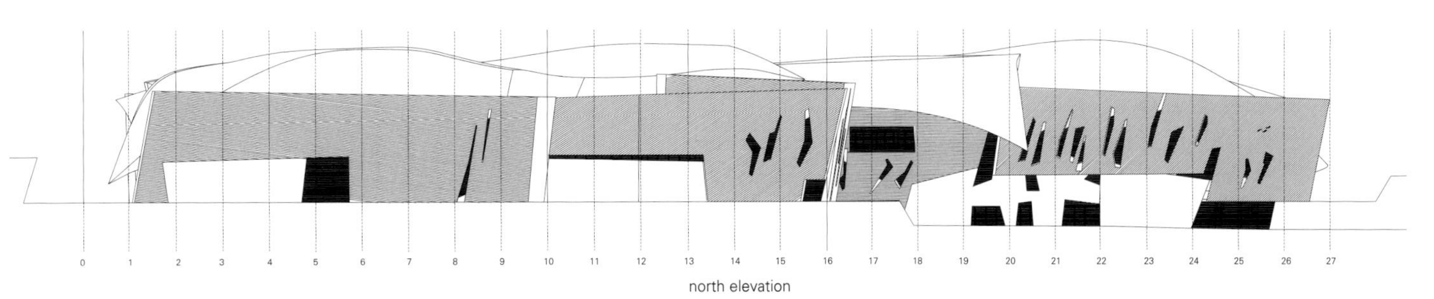

north elevation

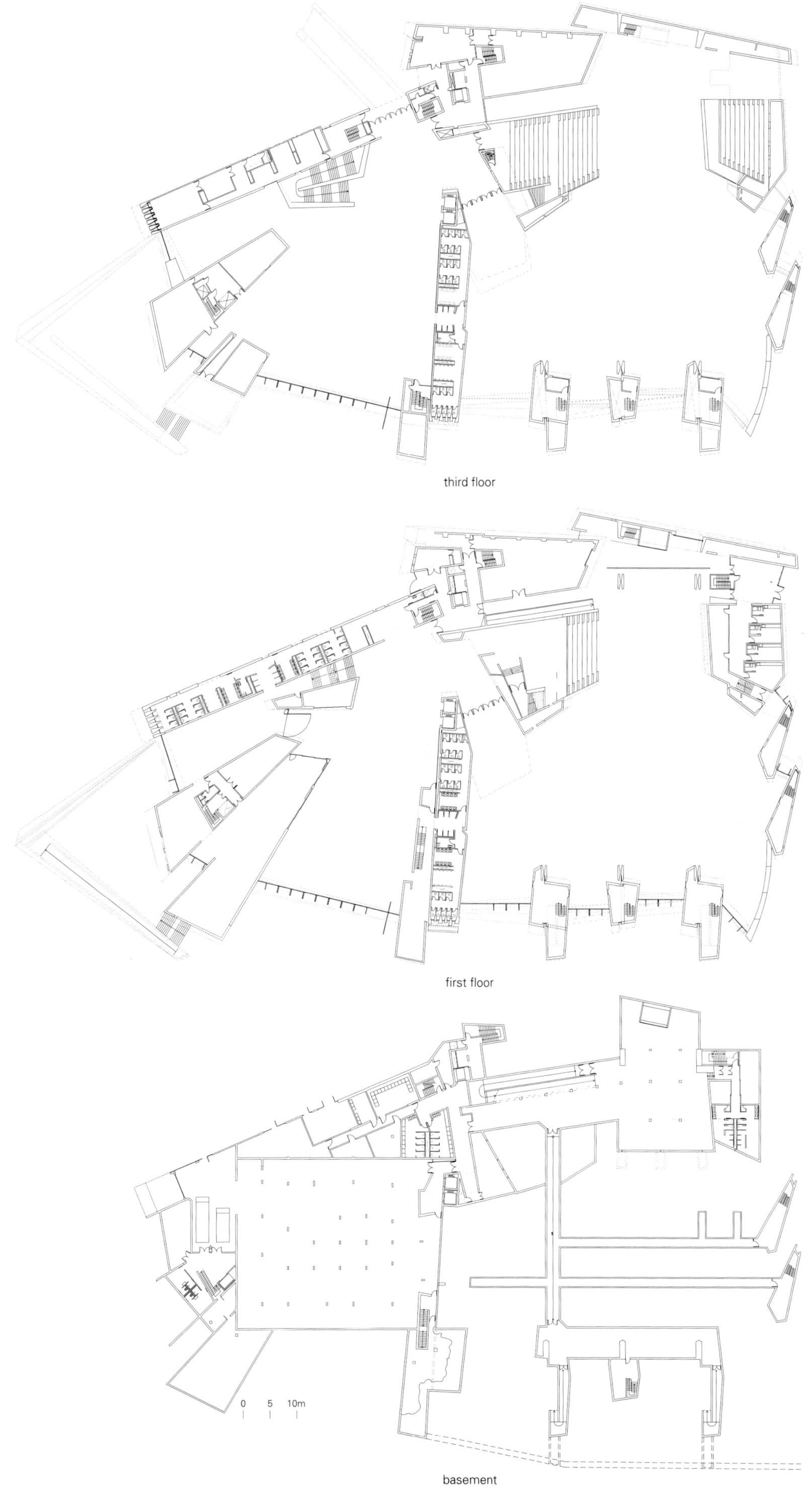

third floor

first floor

basement

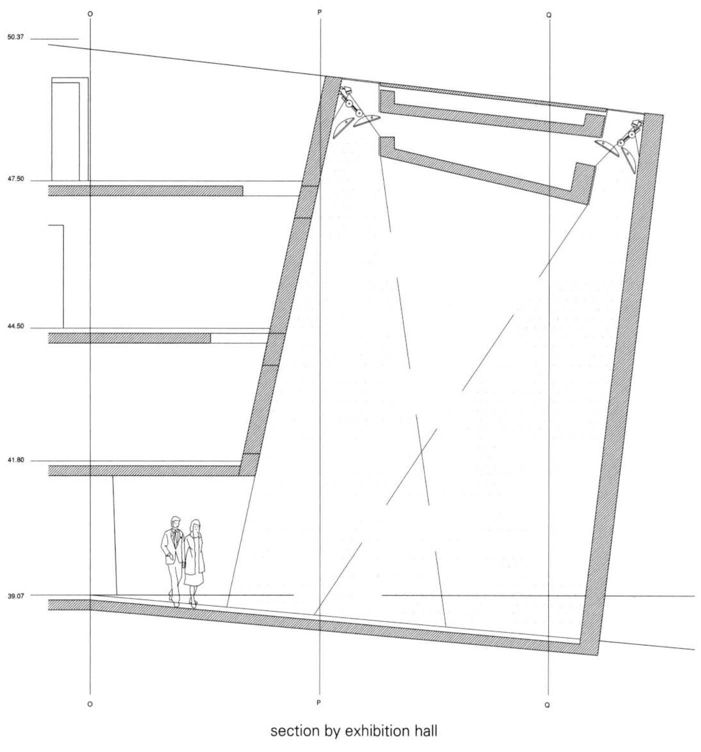

section by exhibition hall

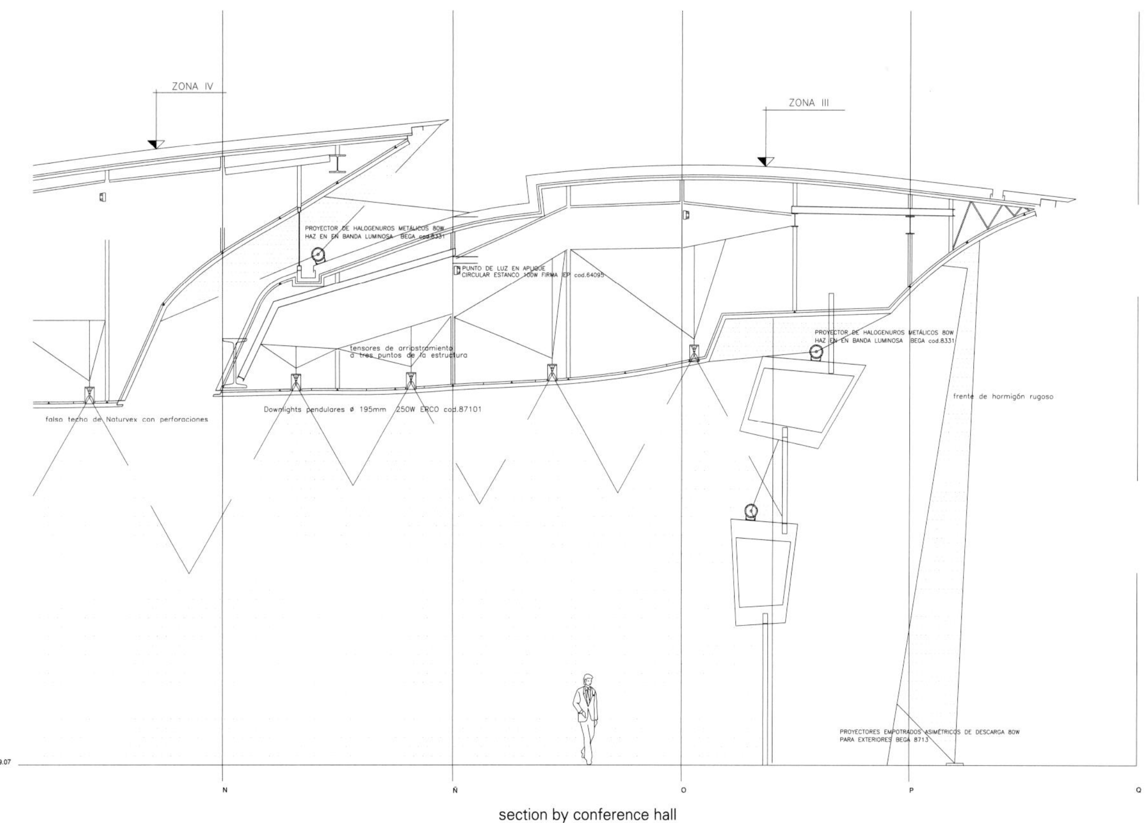

section by conference hall

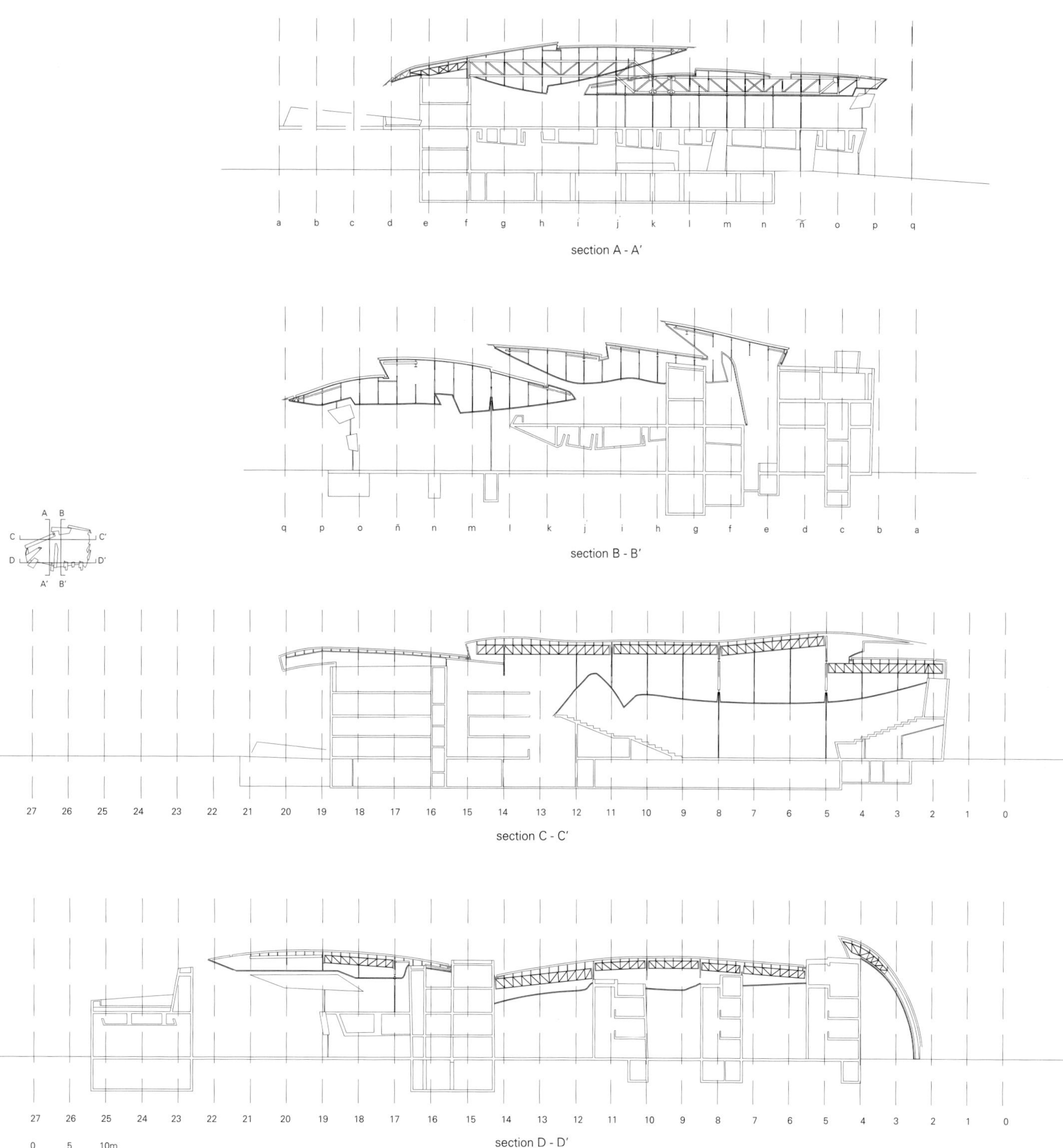

section A - A'

section B - B'

section C - C'

section D - D'

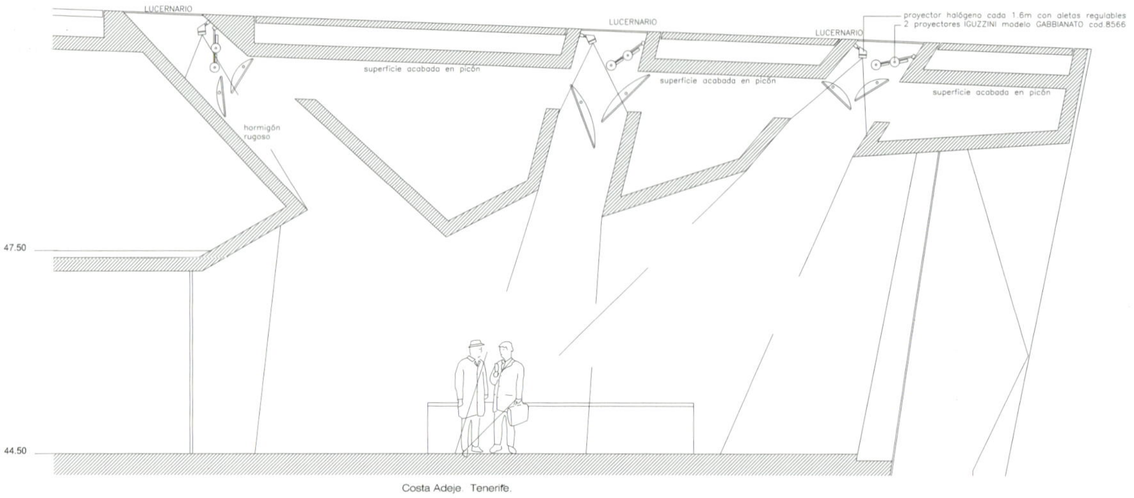

section by office meeting room

The facades are composed of concrete whose texture echoes the decomposition of the Mondariz granite hung from the metal profile and oak carpentries in a T form, with the uppermost part of the granite rising over the profile of the roof. The play of concrete and oak continue into the interior, where the geometry of the stairs connect the levels and spaces which are flooded with the light from above.
The building's program includes the chancellor' and vice-chancellor's offices, the press office, the general secretary and other administrative personnel, and meeting rooms arranged in a radial form. Some spaces open to others and all spectacularly frame views through the structure. Natural light penetrates everything.
The main hall is a ramped auditorium that is projected over the larger spaces between the tripartite structure in a T form, whose top wall is in the opposite direction of the stairs and creates an impressive form on the lateral facade.

비고 대학교 본관

이 건물의 입면은 몬다리스 지방 화강암의 질감을 표현한 콘크리트가 금속 소재와 T자 형 오크목에 매달려 있으며, 가장 높은 부분의 화강암은 지붕 너머로 솟아 올라 있다. 콘크리트와 오크목의 조합은 내부에서도 이어지고 햇빛이 쏟아져내리는 내부의 계단으로 층과 공간이 연결된다.
이 건물에는 총장실과 부총장실, 보도국, 서기실과 기타 관리 인사부, 회의장 등이 방사형으로 배치되어 있다. 어떤 공간들은 다른 공간으로 열려 있고, 어느 곳에서나 멋진 조망을 바라볼 수 있으며 햇빛이 들어온다.
대강당은 세 갈래로 나누어진 T자 형 구조 사이에 있다. 이 T자 형 구조의 최상부는 계단 맞은 편에 있으며 인상적인 입면을 만든다.

Miralles Tagliabue EMBT

Rector Office at Vigo University Campus

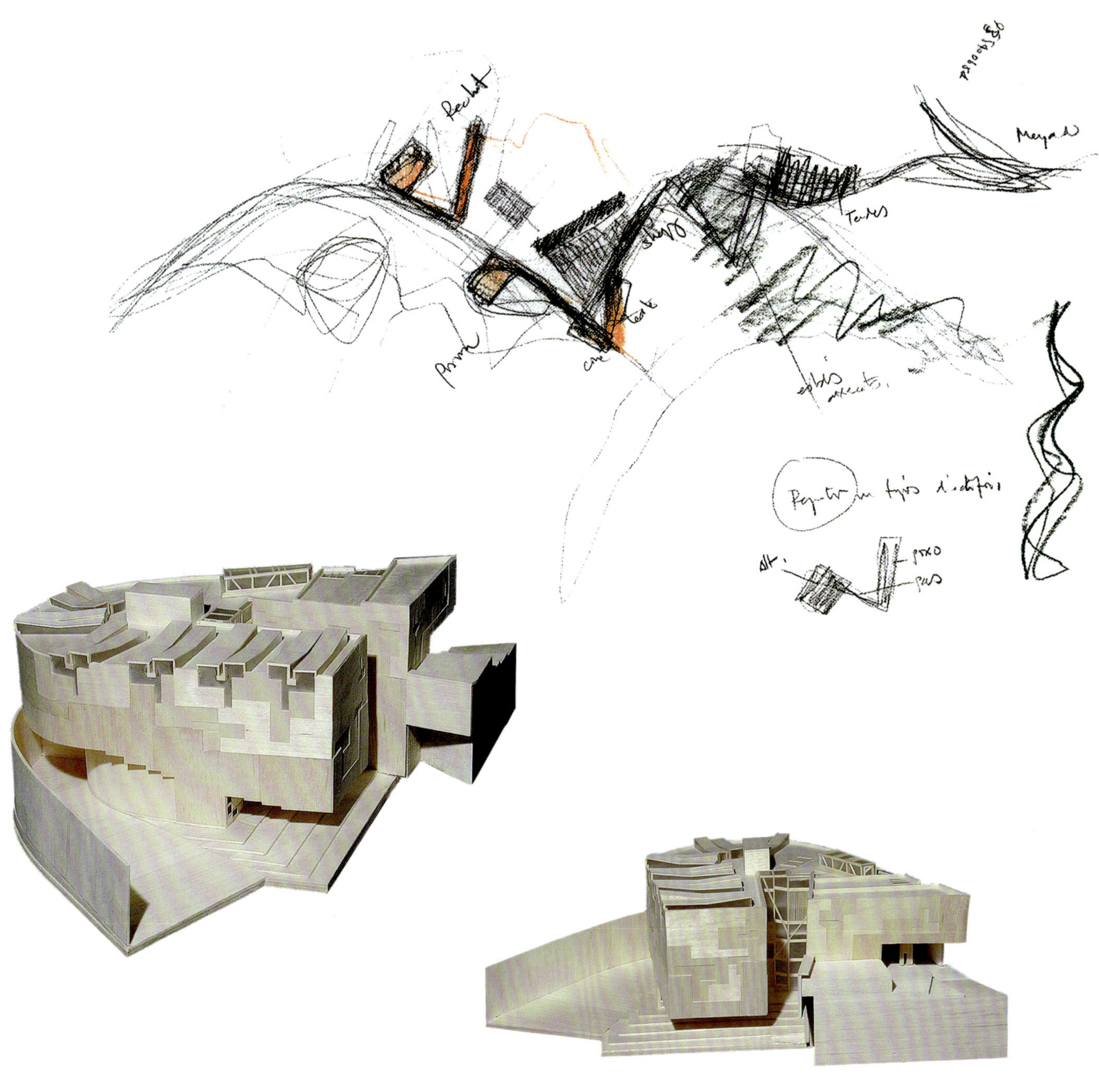

Architect Miralles Tagliabue EMBT- Enric Miralles, Benedetta Tagliabue **Project director** Dani Rossello **Collaborator** Ll. Corbella, E. Nedelcu, N. Laverde, K. Bonifaz, P. Michaud, P. Sándor Nagy, S. Saraiva, J. Bardelli, S. De Vos, S. Nunes Henriques, C. Vernier, J. Villamil, J. Krüguer, A. Quantrill, R. Patti, D. Mayer, D. Erfeld, E. Farkas, G. Rotelli, J.Williams, J. Scheifele, V. Alfonsi, S. Leone, M. de Gregorio, G. Fascioli (model workshop) **Structures** IOC, Nilo Lletjós **Installations** PROISOTEC, Josep Masachs **Technical architect** Manuel Cuquejo Gonzalez, Xose Rodriguez, Tecnics G3 **Constructor** Constructora San José **Safety on site** Julio Millara Turienzo **Location** Vigo, Spain **Detailed design period** 2004~2006 **Photograph** Duccio Malagamba

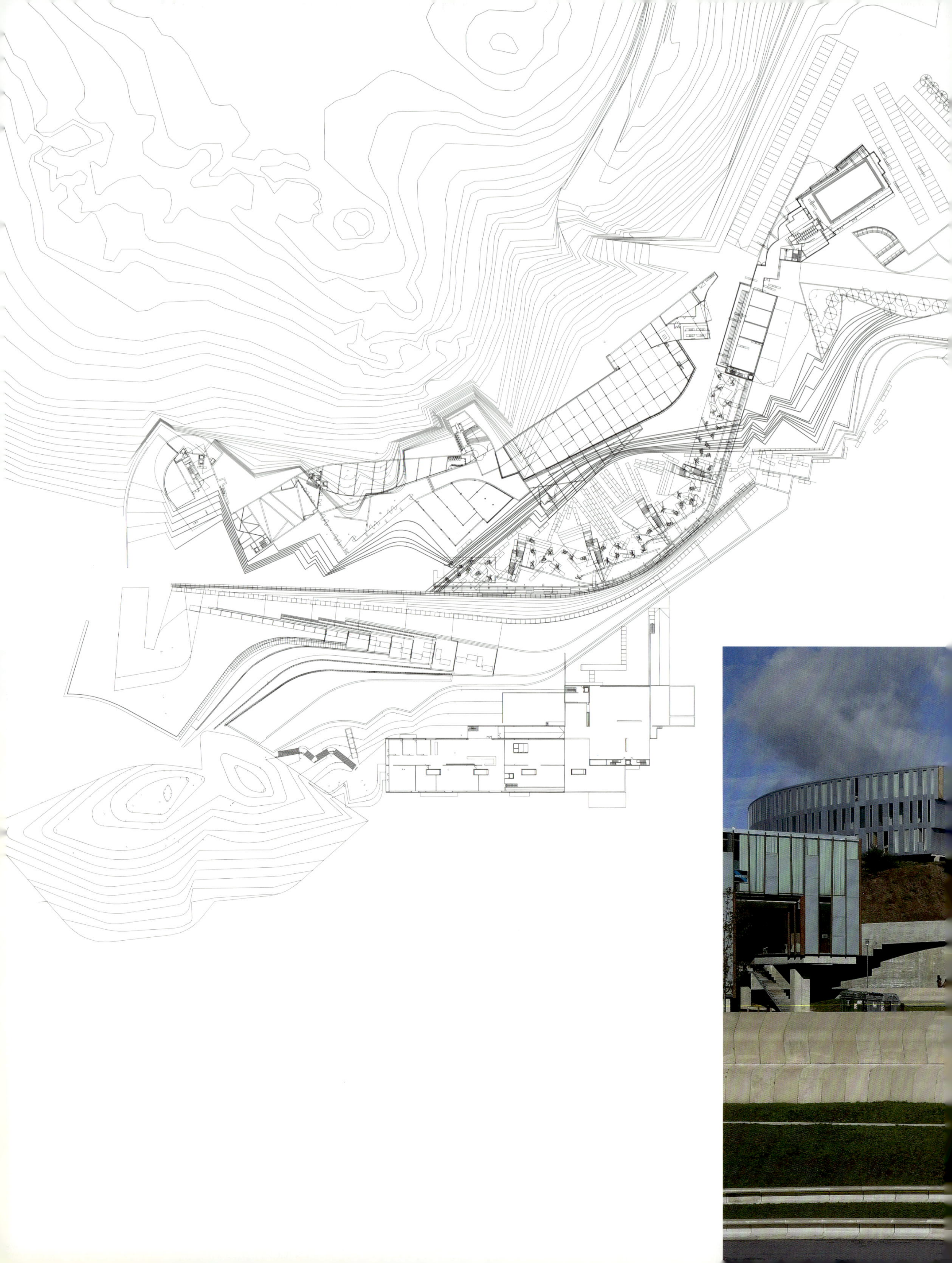

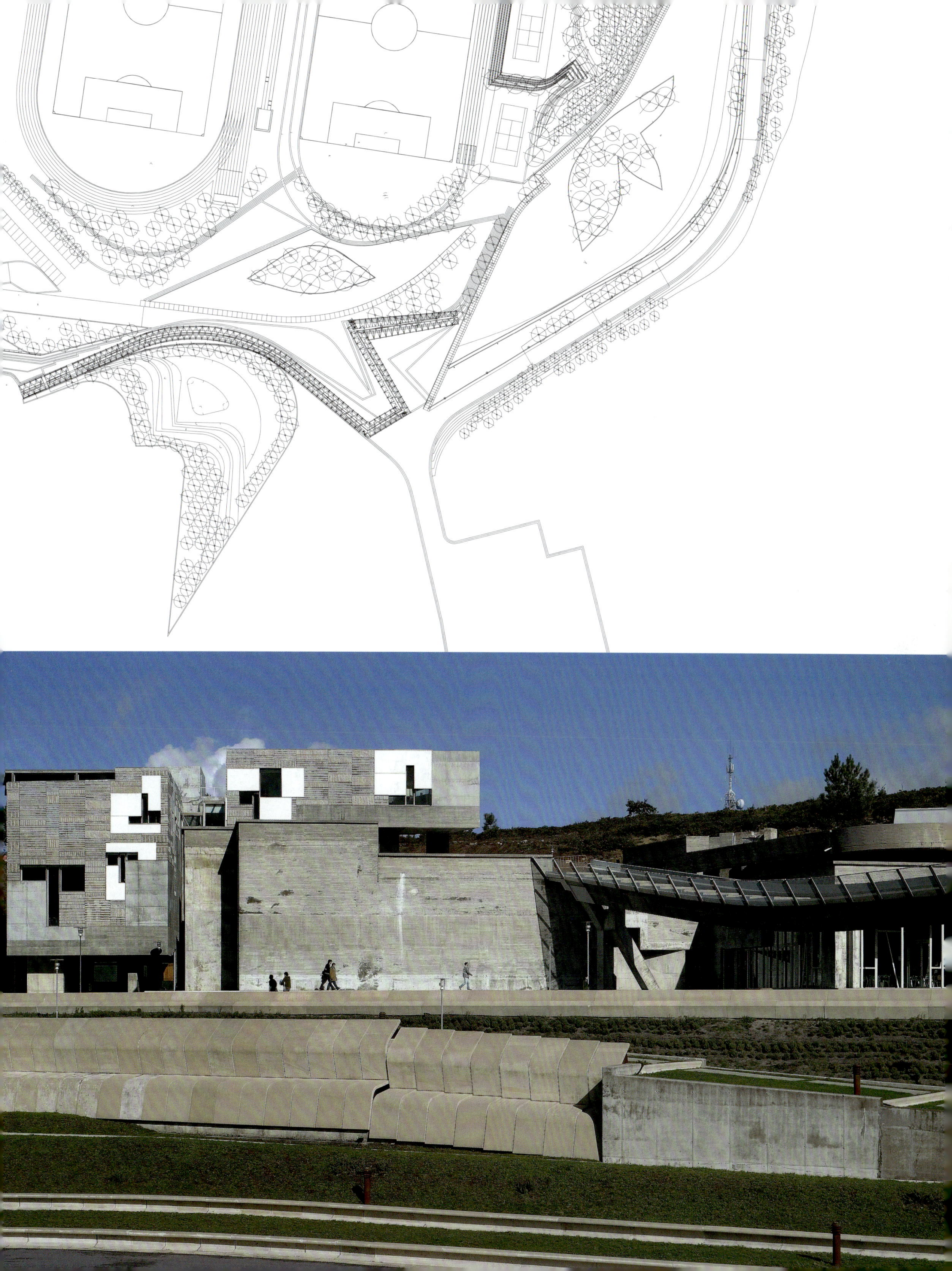

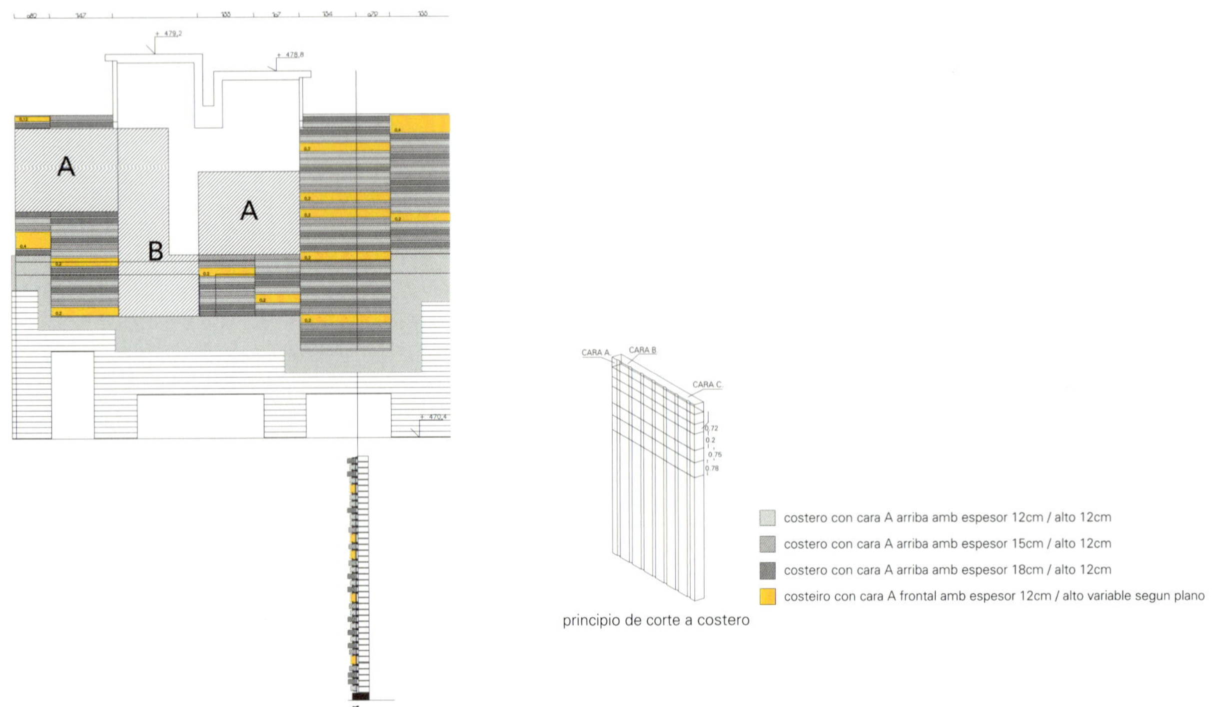
A
A
B
CARA A
CARA B
CARA C
costero con cara A arriba amb espesor 12cm / alto 12cm
costero con cara A arriba amb espesor 15cm / alto 12cm
costero con cara A arriba amb espesor 18cm / alto 12cm
costeiro con cara A frontal amb espesor 12cm / alto variable segun plano
principio de corte a costero

A
B'
A
A
A
A

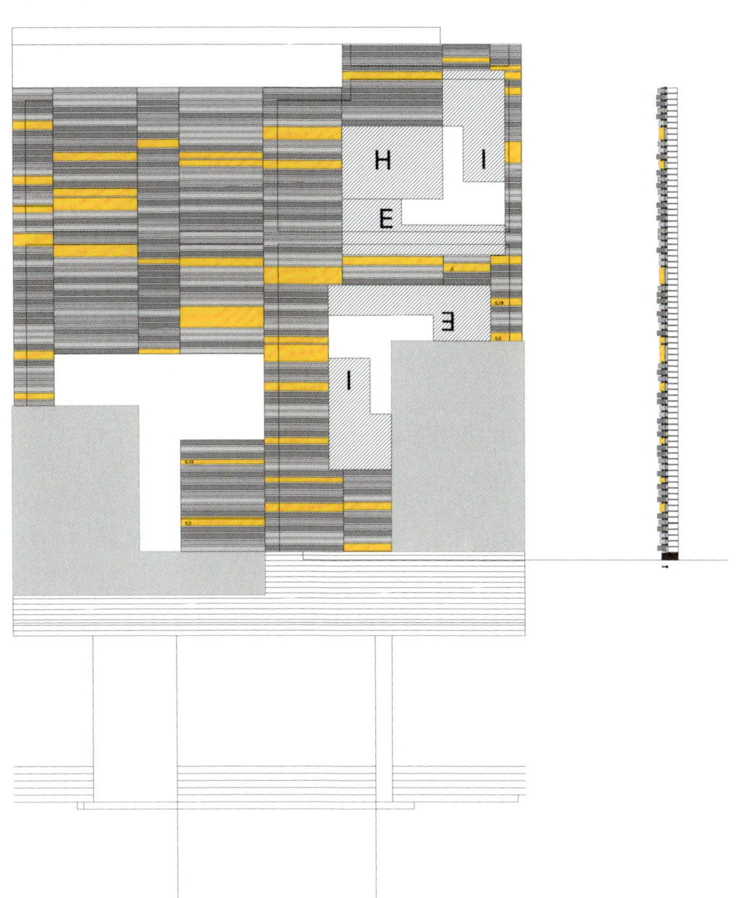
H
I
E
Ǝ
I

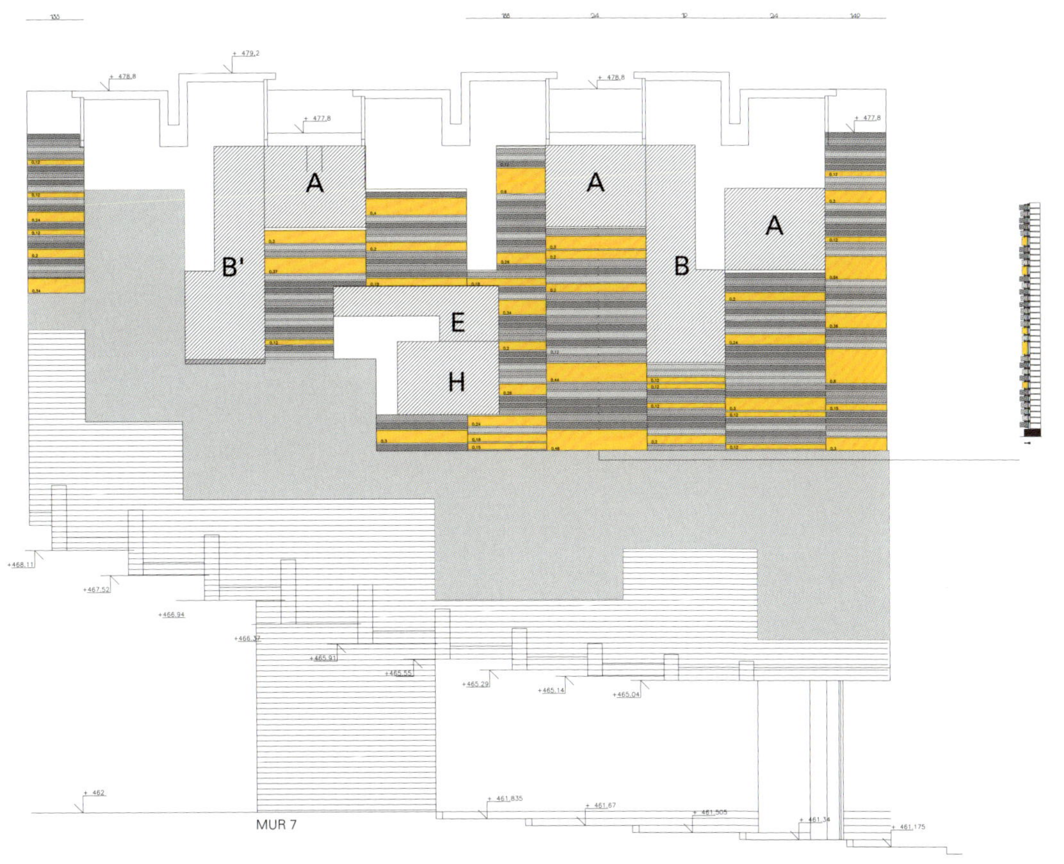
A
A
A
B'
B
E
H
MUR 7

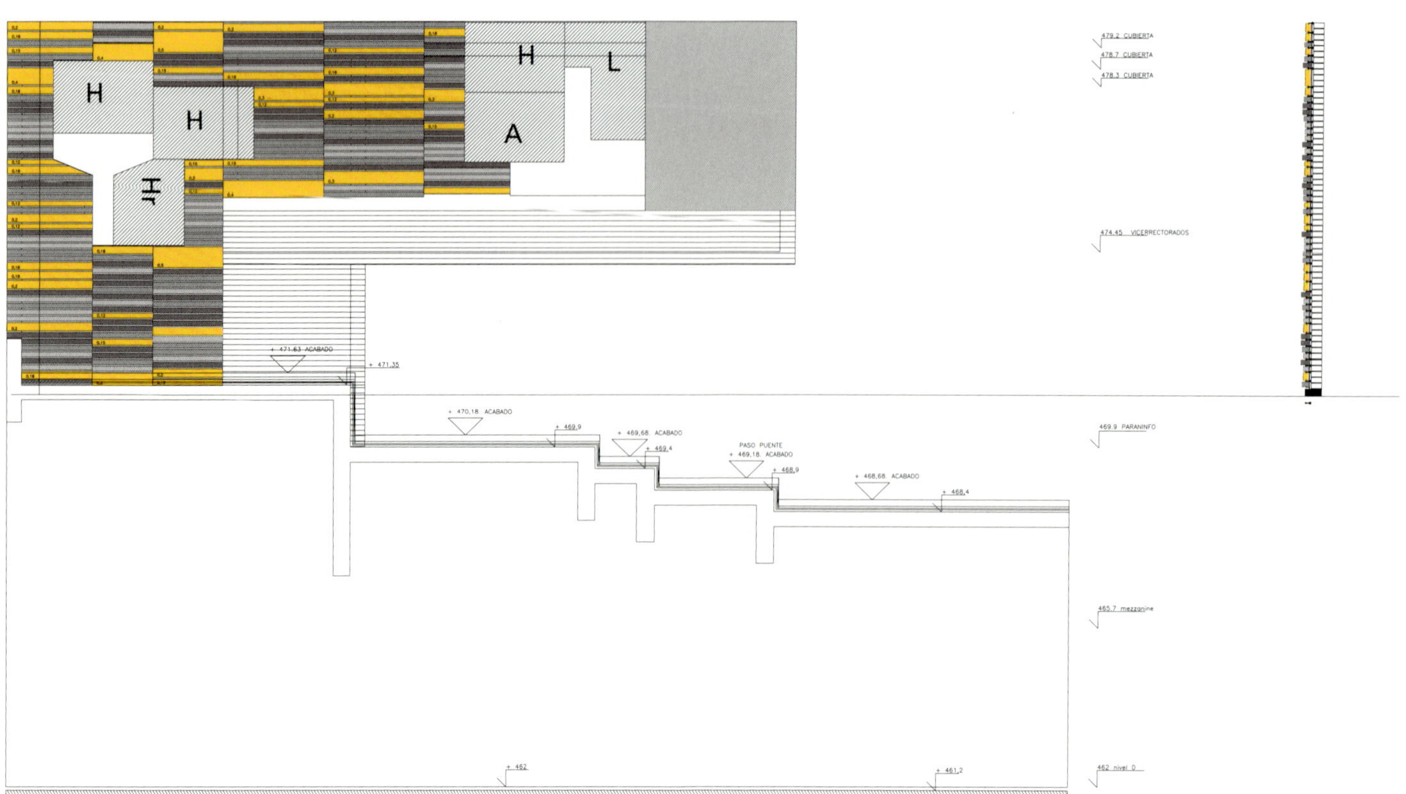
H
H
L
H
A
Hr
479,2 CUBIERTA
478,7 CUBIERTA
478,3 CUBIERTA
474,45 VICERRECTORADOS
+ 471,63 ACABADO
+ 471,35
+ 470,18 ACABADO
+ 469,9
+ 469,68 ACABADO
+ 469,4
PASO PUENTE
+ 469,18 ACABADO
+ 468,9
+ 468,68 ACABADO
+ 468,4
469,9 PARANINFO
465,7 mezzanine
+ 462
+ 461,2
462 nivel 0

MUR 11

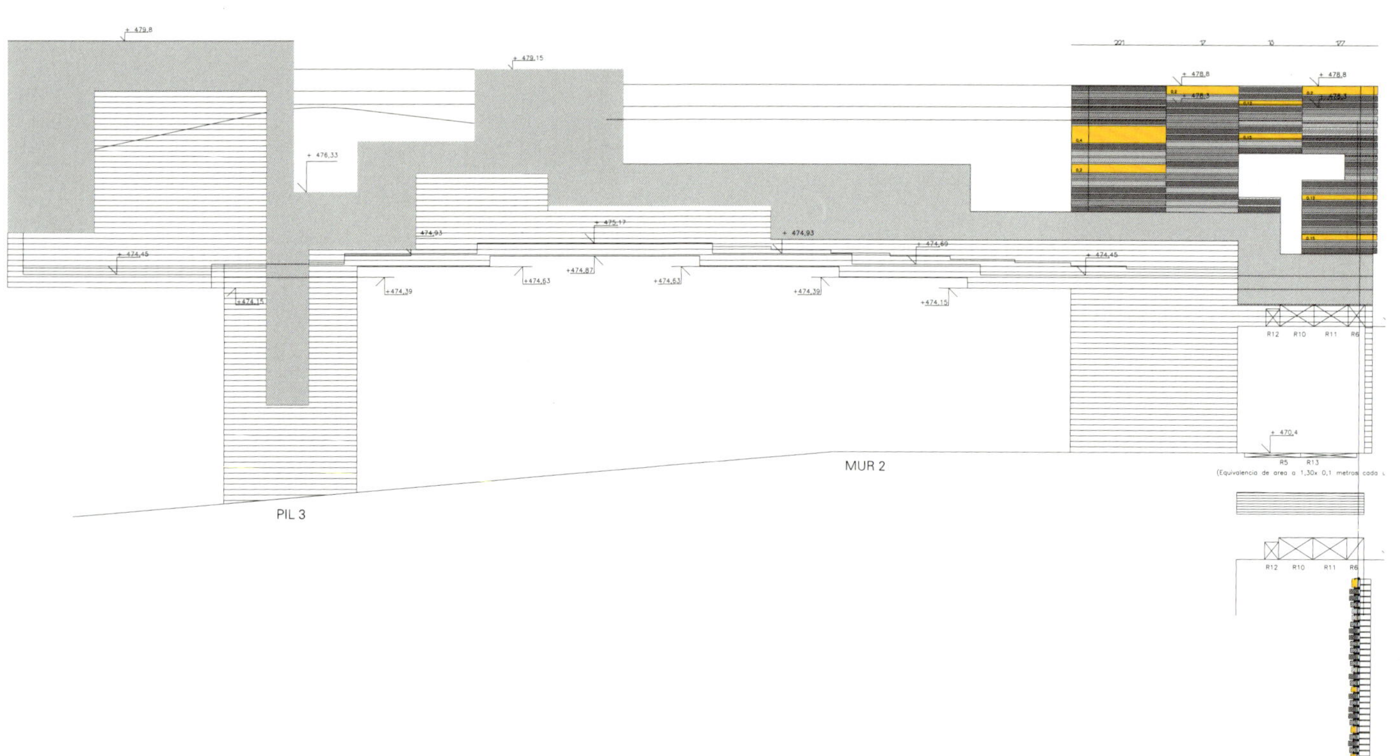
+ 479,8
+ 479,15
+ 476,33
+ 478,8
+ 478,3
+ 474,93
+ 475,17
+ 474,45
+ 474,69
+474,39
+474,63
+474,87
+474,15
R12 R10 R11 R6
+ 470,4
R5 R13
MUR 2
PIL 3

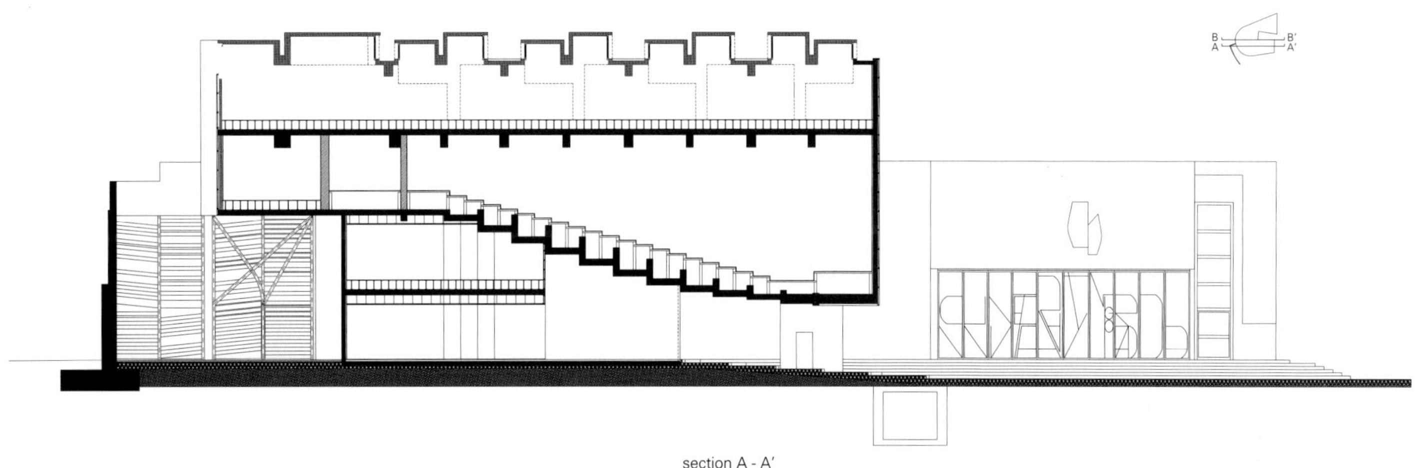

section A - A′

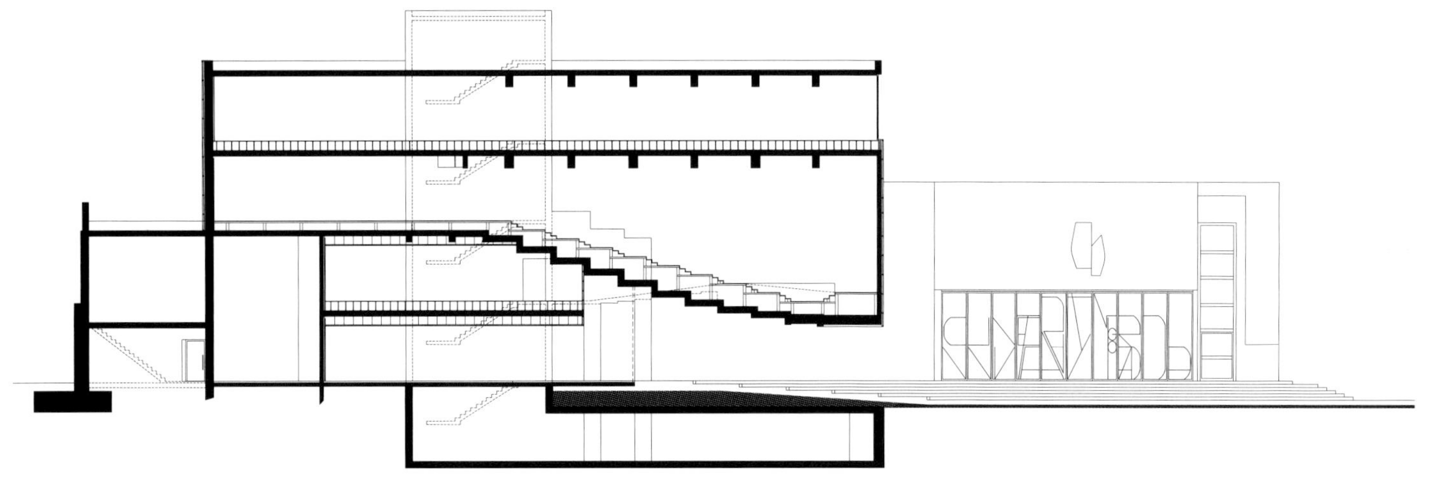

section B - B′

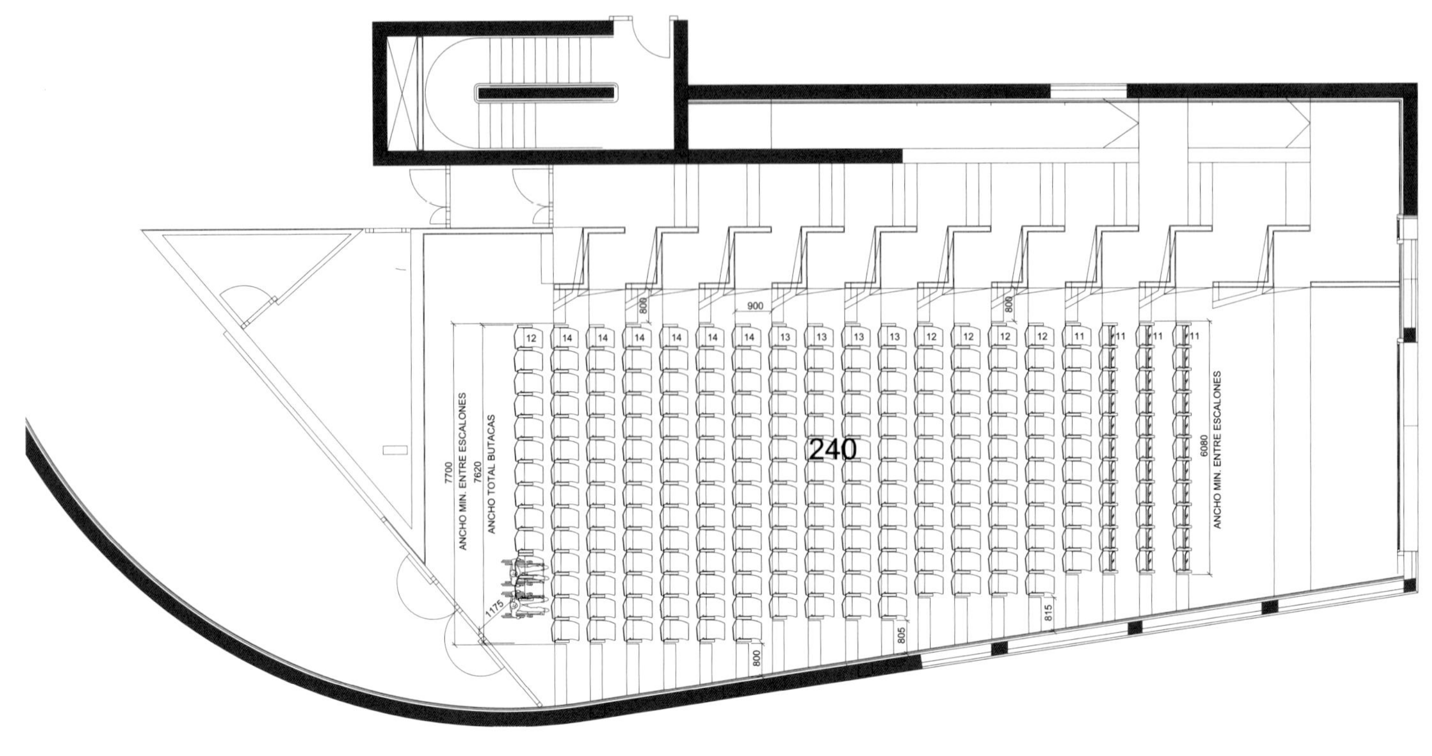
240
7700
ANCHO MIN. ENTRE ESCALONES
7620
ANCHO TOTAL BUTACAS
6080
ANCHO MIN. ENTRE ESCALONES
900
800
1175
805
815

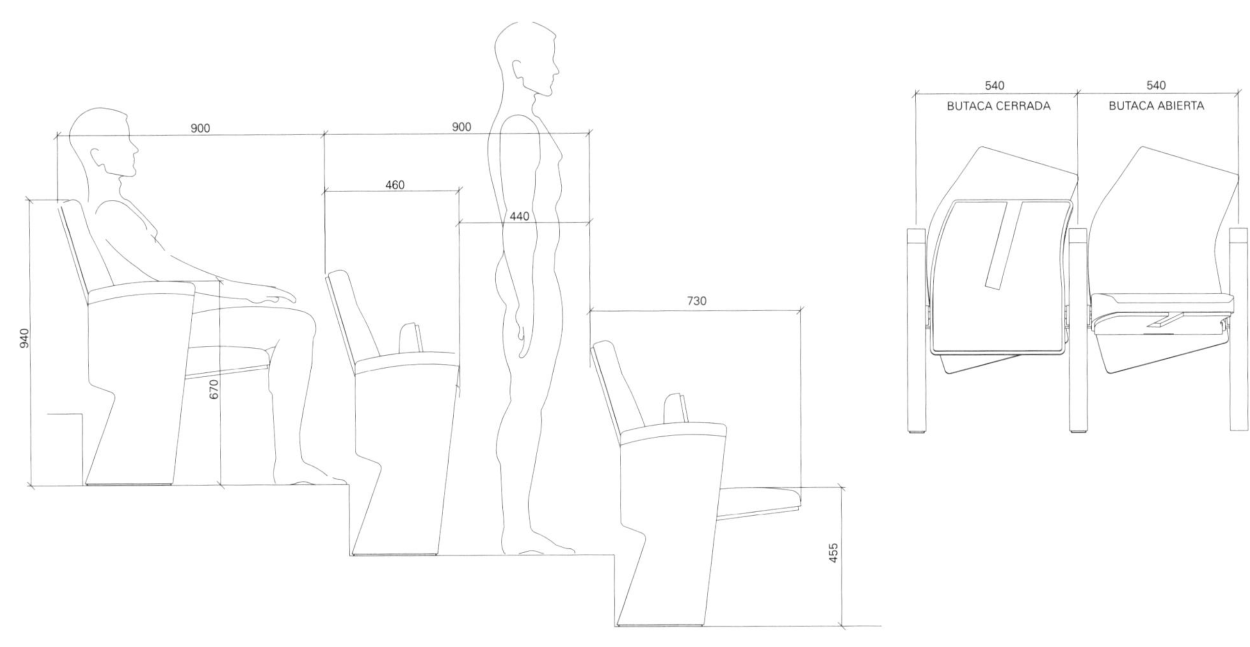
900
900
460
440
730
940
670
455
540
540
BUTACA CERRADA
BUTACA ABIERTA

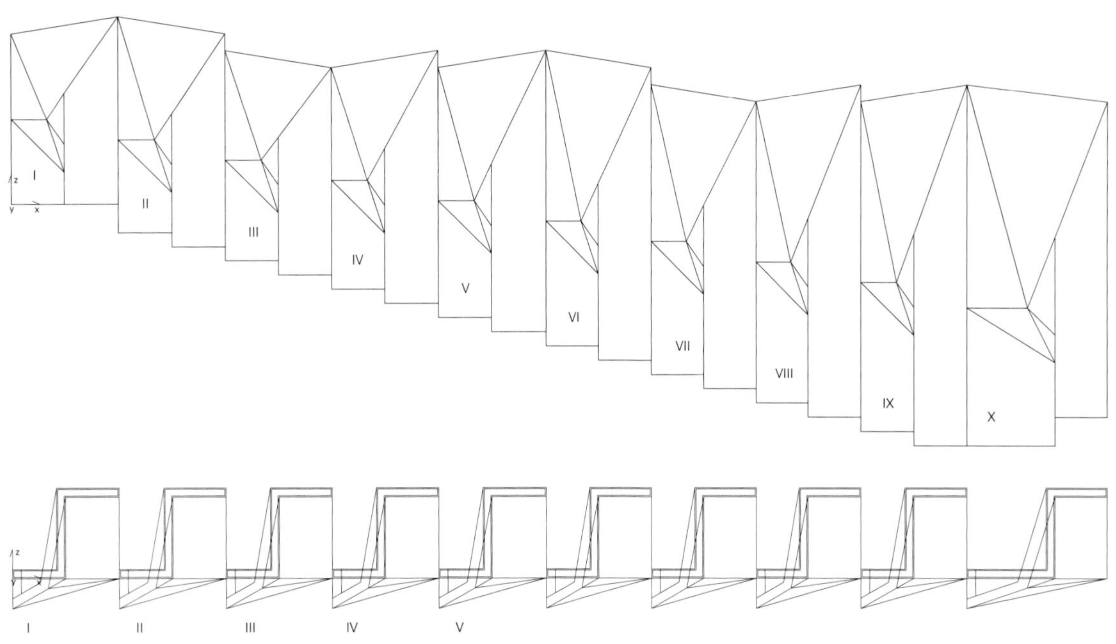
I
II
III
IV
V
VI
VII
VIII
IX
X
I
II
III
IV
V

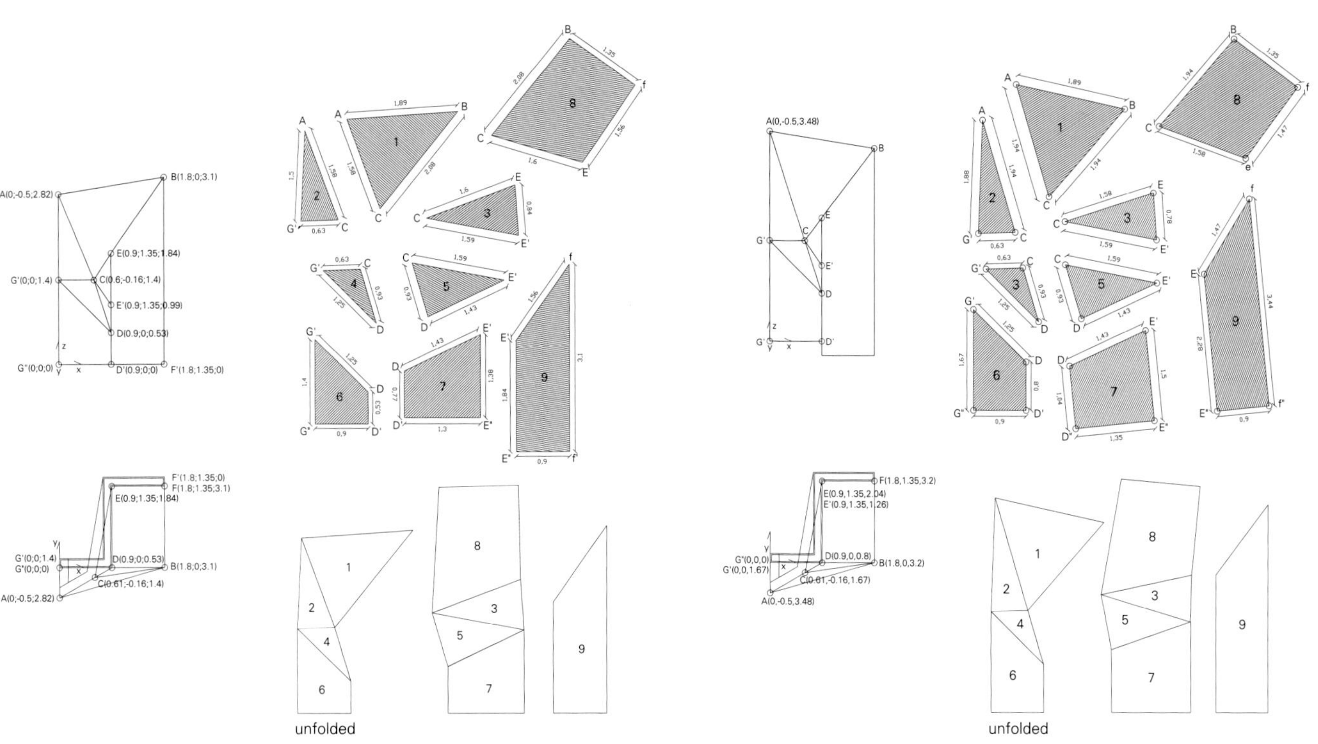
A(0;-0.5;2.82)
B(1.8;0;3.1)
E(0.9;1.35;1.84)
G'(0;0;1.4)
C(0.6;-0.16;1.4)
E'(0.9;1.35;0.99)
D(0.9;0;0.53)
G''(0;0;0)
D'(0.9;0;0)
F'(1.8;1.35;0)
F'(1.8;1.35;0)
F(1.8;1.35;3.1)
G'(0;0;1.4)
G''(0;0;0)
D(0.9;0;0.53)
B(1.8;0;3.1)
C(0.61;-0.16;1.4)
A(0;-0.5;2.82)
unfolded
A(0,-0.5,3.48)
F(1.8,1.35,3.2)
E(0.9,1.35,2.04)
E'(0.9,1.35,1.26)
G''(0,0,0)
G'(0,0,1.67)
D(0.9,0,0.8)
B(1.8,0,3.2)
C(0.61,-0.16,1.67)
A(0,-0.5,3.48)
unfolded

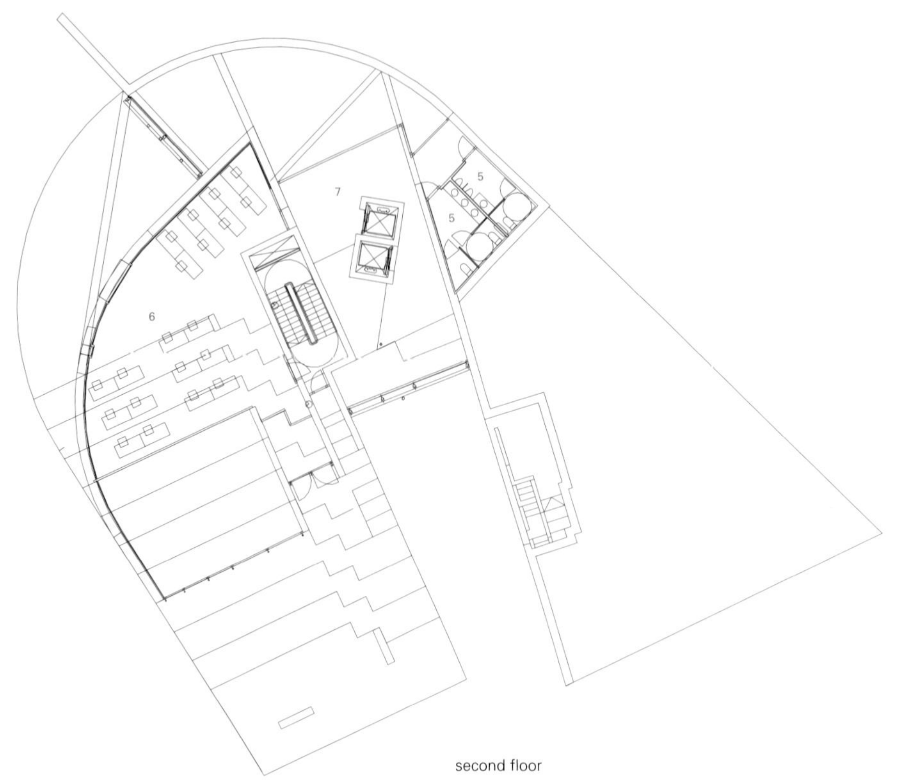

second floor

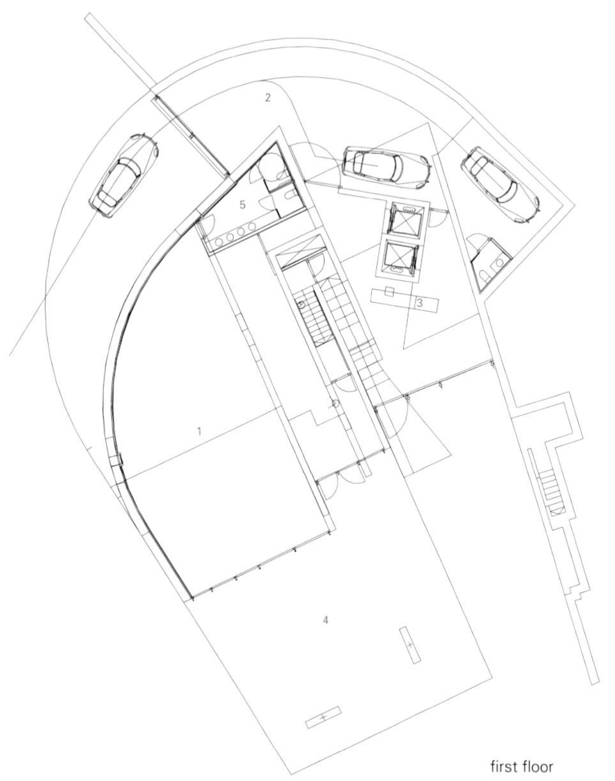

first floor

1. children space
2. parking
3. entrance
4. outer seats
5. toilet
6. press office / legal advice office
7. communication space
8. auditorium
9. access auditorium
10. hall
11. general secretary
12. projection room
13. vice-chancellor's office
14. meeting room
15. vice-chancellor's secretary
16. secretary zone
17. chancellor's office
18. chancellor's meeting room
19. chancellor's secretary

fourth floor

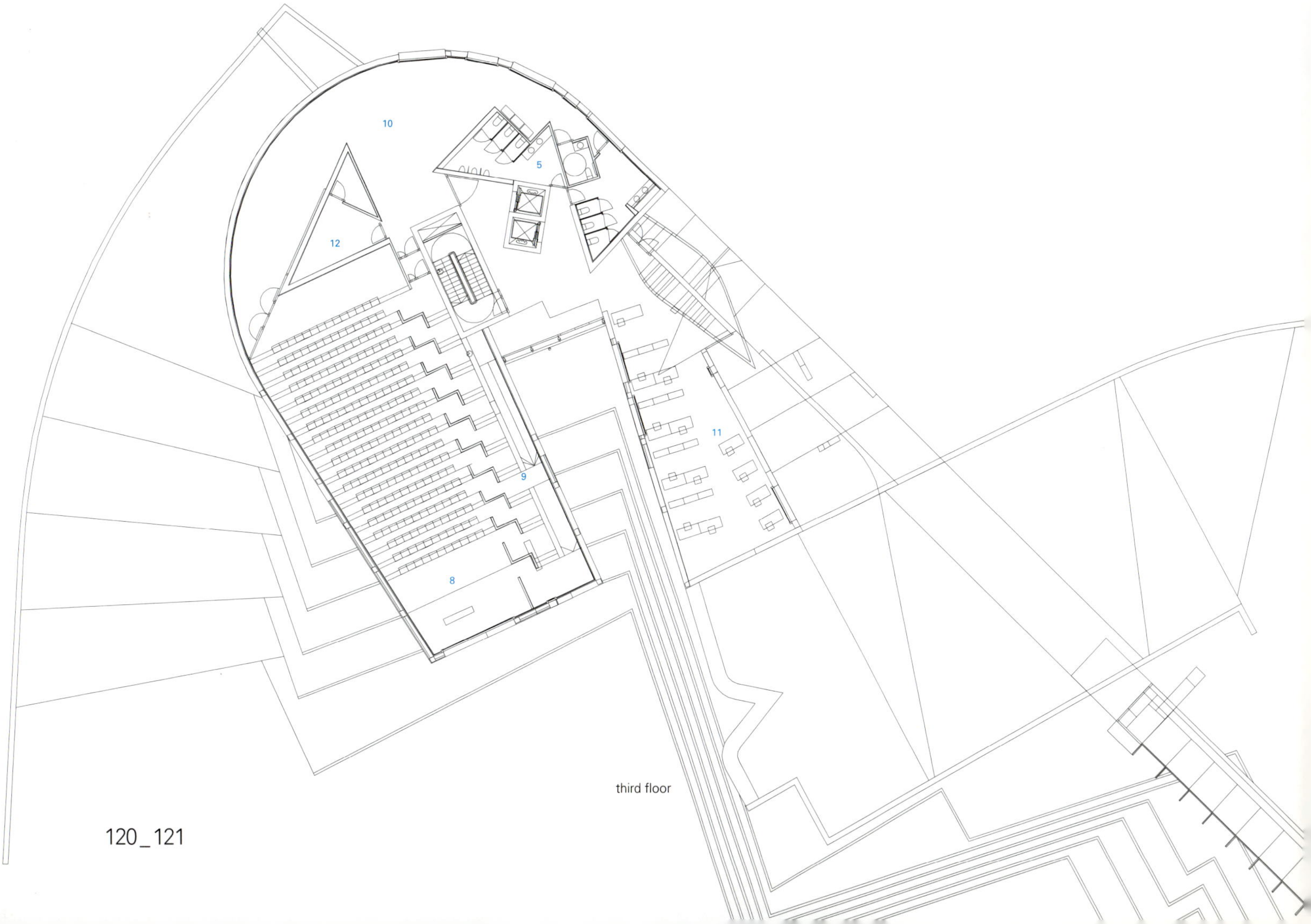

third floor

A Magnet

The CaixaForum is conceived as an urban magnet attracting not only the art-lovers but all people of Madrid and from outside. The attraction will not only be CaixaForum's cultural program, but also the building itself, insofar that its heavy mass is detached from the ground in apparent defiance of the laws of gravity and, in a real sense, draw the visitors inside.

A new address for the arts

The CaixaForum-Madrid stands on an advantageous site facing the Paseo del Prado and the Botanical Garden vis a vis. This new address for the arts is located in an area occupied until now by unspectacular urban structures, the Central Electrica Power Station, and a gas station. The classified brick walls of the former power station are reminiscences of the early industrial age in Madrid, while the gas station, a purely functional structure, was clearly out of place. Like a vineyard that could never develop its full potential because it was planted with the wrong grape, this prominent location could not develop its full potential. The demolition of the gas station created a small plaza between the Paseo del Prado and the new CaixaForum in the converted power station.

A spectacular transformation

The only material of the old power station that we could use was the classified brick shell. In order to conceive and insert the new architectural components of the CaixaForum Project, we

까익샤포룸

상징성

까익샤포룸은 예술애호가뿐만 아니라 마드리드를 사랑하는 모든 사람들에게 매력적인 도시의 상징물로 각인된다. 이것은 까익샤포룸의 문화 프로그램과 더불어 건물 자체의 매력에서 비롯된 것이다. 중력에 도전하듯 지면에서 떠있는 거대한 부피의 매스가 방문자를 그 안으로 이끈다.

예술의 새로운 획을 긋다

마드리드의 까익샤포룸은 파세오 델 프라도 가와 식물원을 마주하고 있어 입지 조건이 좋다. 이 지역은 지금까지 평범한 도시 건물들과 중앙 발전소 그리고 주유소가 차지하고 있었다. 예전 발전소로 쓰였음을 가늠케 하는 오래된 벽돌담은 마드리드 초기 산업사회를 떠오르게 하는 유물이며, 주유소라는 완전히 기능적인 구조물은 이곳에는 어울리지 않았다. 마치 맞지 않는 품종의 포도가 심어져 최상의 품질을 얻을 수 없는 포도원처럼, 이 특별한 지역은 그 무한한 잠재성을 잘 발휘할 수 없었다. 폐허가 되어버린 주유소는 파세오 델 프라도 가와 발전소를 새롭게 탈바꿈한 까익샤포룸 사이의 작은 광장으로 거듭났다.

Herzog & de Meuron

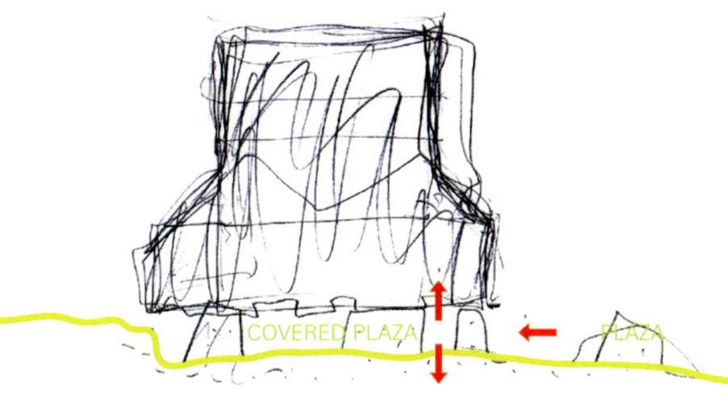

began with a surgical operation, separating and removing the base and the parts of the building no longer needed. This opened a completely novel and spectacular perspective that simultaneously solved a number of problems posed by the site. The removal of the base of the building left a covered plaza under the brick shell, which now appears to float above the street level. This sheltered space under the CaixaForum offers its shade to visitors who want to spend time or meet outside and is at the same time the entrance to the Forum itself. Problems such as the narrowness of the surrounding streets, the placement of the main entrance, and the architectural identity of this contemporary art institution could be addressed and solved in a single urbanistic and sculptural gesture.

A construction below and a construction above ground

The separation of the structure from the ground level creates two worlds: one below and the other above the ground. The 'underworld' buried beneath the topographically landscaped plaza provides space for a theater/auditorium, service rooms, and several parking spaces. The multi-storied building above ground houses the entrance lobby and galleries, a restaurant and administrative offices. There is a contrast between the flexible and loft-like character of the exhibition spaces and the spatial complexity of the top floor with its restaurant/bar and the offices. The surprising sculptural aspect of the CaixaForum's silhouette is no mere architectural fancy, but reflects the roofscape of the surrounding buildings.

눈부신 전환

옛 발전소에서 가져온 재료는 외부에 쓰인 벽돌뿐이다. 새로운 건축적 요소를 구상하여 집어 넣고, 더 이상 필요 없어진 기존 건물의 기초와 일부를 분리하고 제거하기 시작했다. 이로써 대지에 산재해 있는 여러 문제를 해결하는 동시에 완결된 이야기와 풍부한 경관을 선사하게 되었다. 건물의 기단부는 보도면 위로 띄워 벽돌 건물 아래에 두고 그 아래는 광장으로 덮었다. 건물 하부의 공간은 방문객에게 머무름이나 만남을 위한 그늘을 제공하는 동시에 포룸의 입구 역할도 한다. 주변 보도의 좁은 폭, 주 출입구의 배치 그리고 현대 예술센터의 건축적 정체성 등의 문제가 도시적이며 조형적인 태도로 간단히 해결되었다.

지하 건축물과 지상 건축물

이 건물은 지면을 기준으로 위와 아래의 두 세계가 공존한다. 지하 세계에는 극장과 대강당, 서비스실과 주차공간이 있는 광장이 숨 쉬고 있으며, 위층에는 로비와 갤러리, 식당과 관리사무실이 자리하고 있다. 이곳에는 전시 성격에 따라 유연하게 대처할 수 있는 공간과 더불어 꼭대기 층의 식당, 바 그리고 사무실과 같은 복합공간이 잘 대비되고 있다. 까익샤포룸의 눈부신 조형 언어는 단순히 건축적인 환상에 머무르지 않고, 건물을 에워싼 주변의 '루프스케이프'를 반영한다.

Architect Jacques Herzog, Pierre de Meuron, Harry Gugger **Project team** Peter Ferretto, Carlos Gerhard, Stefan Marbach, Benito Blanco, Heitor Garcia Lantaron, Estelle Grosberg, Pedro Guedes, Michel Kehl, Miguel Marcelino, Gabi Mazza, Beatrice Noves Salto, Margarita Salmeron, Stefano Tagliacarne **Associate architect** Mateu i Bausells Arquitectura **Project management** Servihabitat, Construccion i Control **General contractor** Ferrovial Agroman **Structural engineer** WGG Schnetzer Puskas Ingenieure, NB35 **MEP Engineer** Urculo Ingenieros **Facade consulting** Emmer Pfenninger Partner AG, ENAR **Lighting** Arup Lighting **Acoustics** Audioscan **Green wall** Herzog & de Meuron in collaboration with Patrick Blanc, Artist-Botanist **Green wall consultant** Benavides & Lapèrche **Client** Obra Social Fundacion "LaCaixa", Caixa d'Estalvis i Pensions de Barcelona **Location** Madrid, Spain **Site area** building site - 1,934 m², plaza - 650 m² **Bldg. area** 1,400 m² **Gross floor area** 11,000 m² **Design** 2001~2003 **Construction** 2003~2008 **Photograph** Iwan Baan
all material provided ©Herzog & de Meuron

HOTEL

CaixaForum

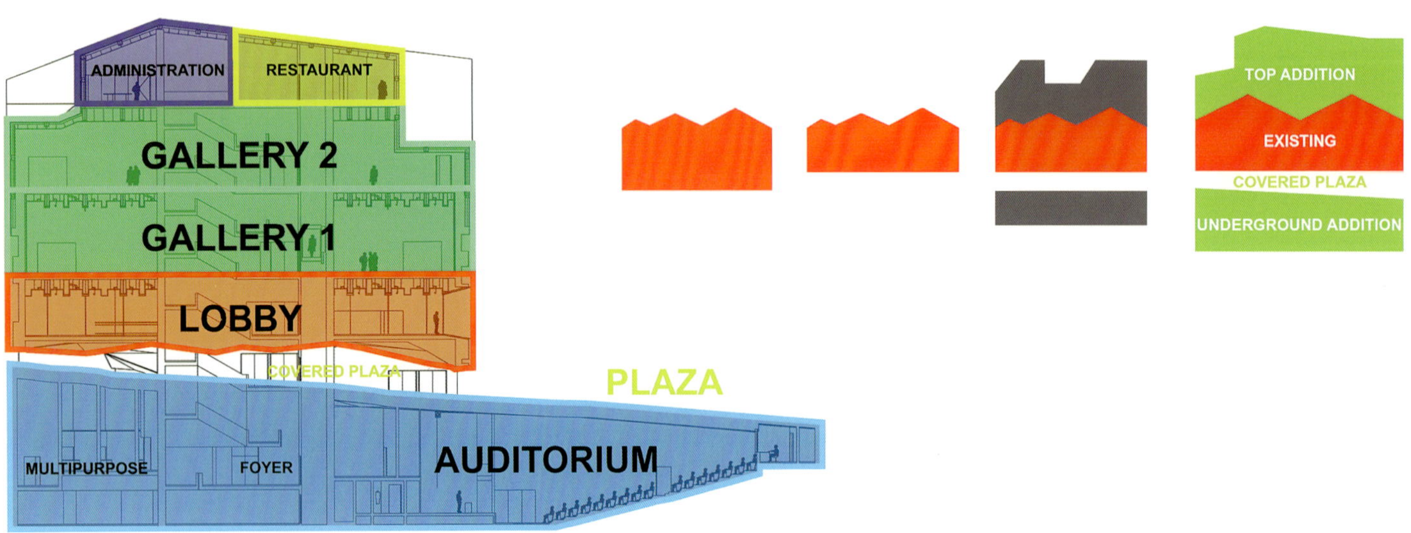
ADMINISTRATION
RESTAURANT
GALLERY 2
GALLERY 1
LOBBY
COVERED PLAZA
PLAZA
MULTIPURPOSE
FOYER
AUDITORIUM
TOP ADDITION
EXISTING
COVERED PLAZA
UNDERGROUND ADDITION

CENTRO

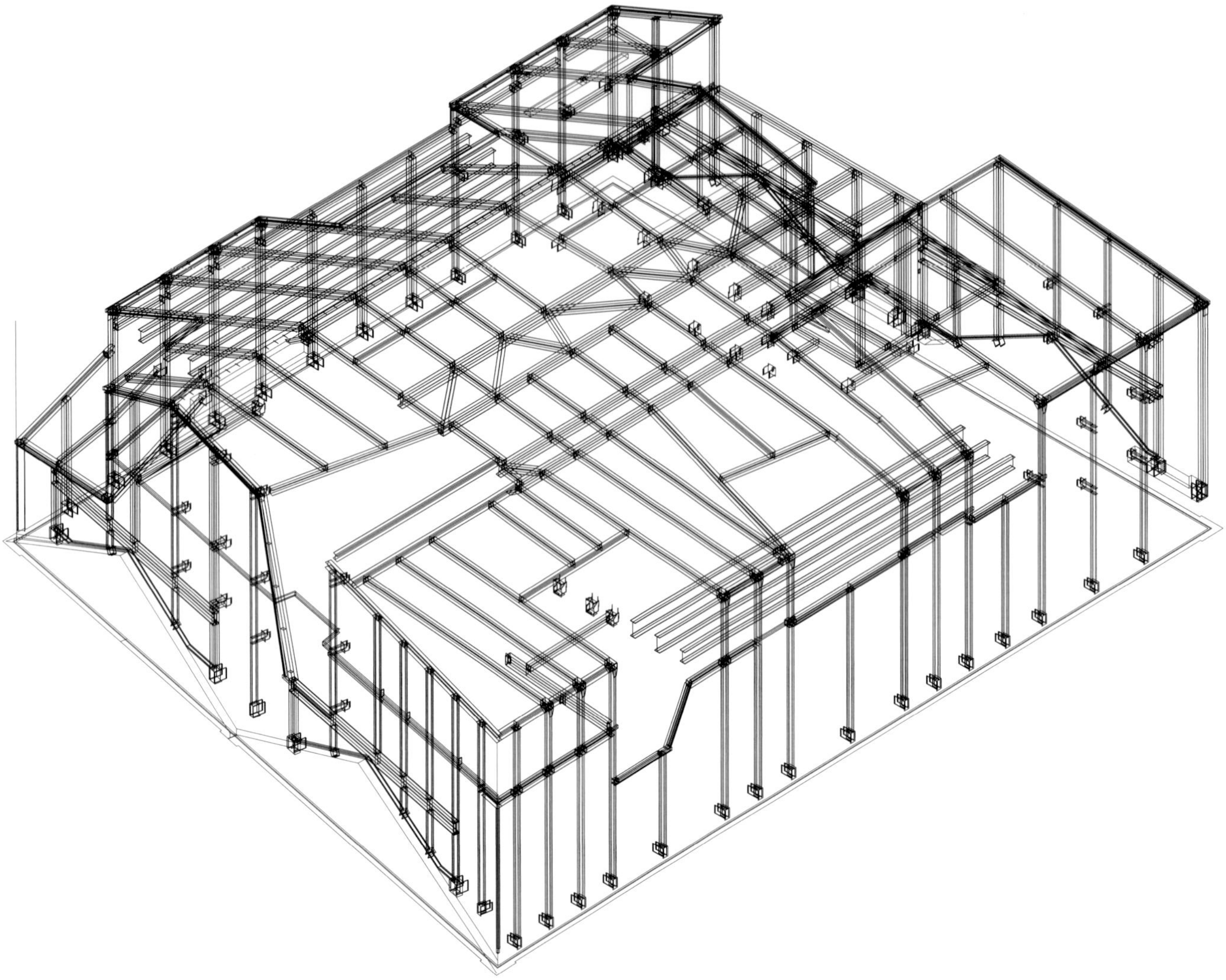

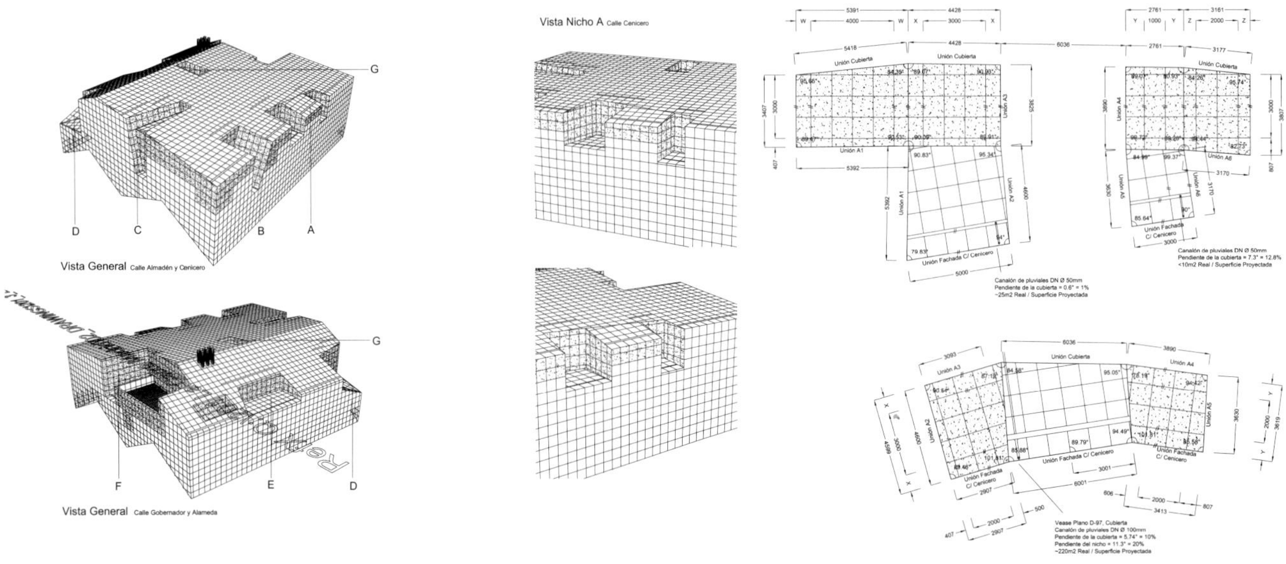
G
D
C
B
A
Vista General Calle Almadén y Cenicero
G
F
E
D
Vista General Calle Gobernador y Alameda
Vista Nicho A Calle Cenicero
Unión Cubierta
Unión A1
Unión A2
Unión A3
Unión Fachada C/ Cenicero
Canalón de pluviales DN Ø 50mm
Pendiente de la cubierta = 0.6° = 1%
~25m2 Real / Superficie Proyectada
Unión A4
Unión A5
Unión A6
Unión Fachada C/ Cenicero
Canalón de pluviales DN Ø 50mm
Pendiente de la cubierta = 7.3° = 12.8%
<10m2 Real / Superficie Proyectada
Vease Plano D-97, Cubierta
Canalón de pluviales DN Ø 100mm
Pendiente de la cubierta = 5.74° = 10%
Pendiente del nicho = 11.3° = 20%
~220m2 Real / Superficie Proyectada

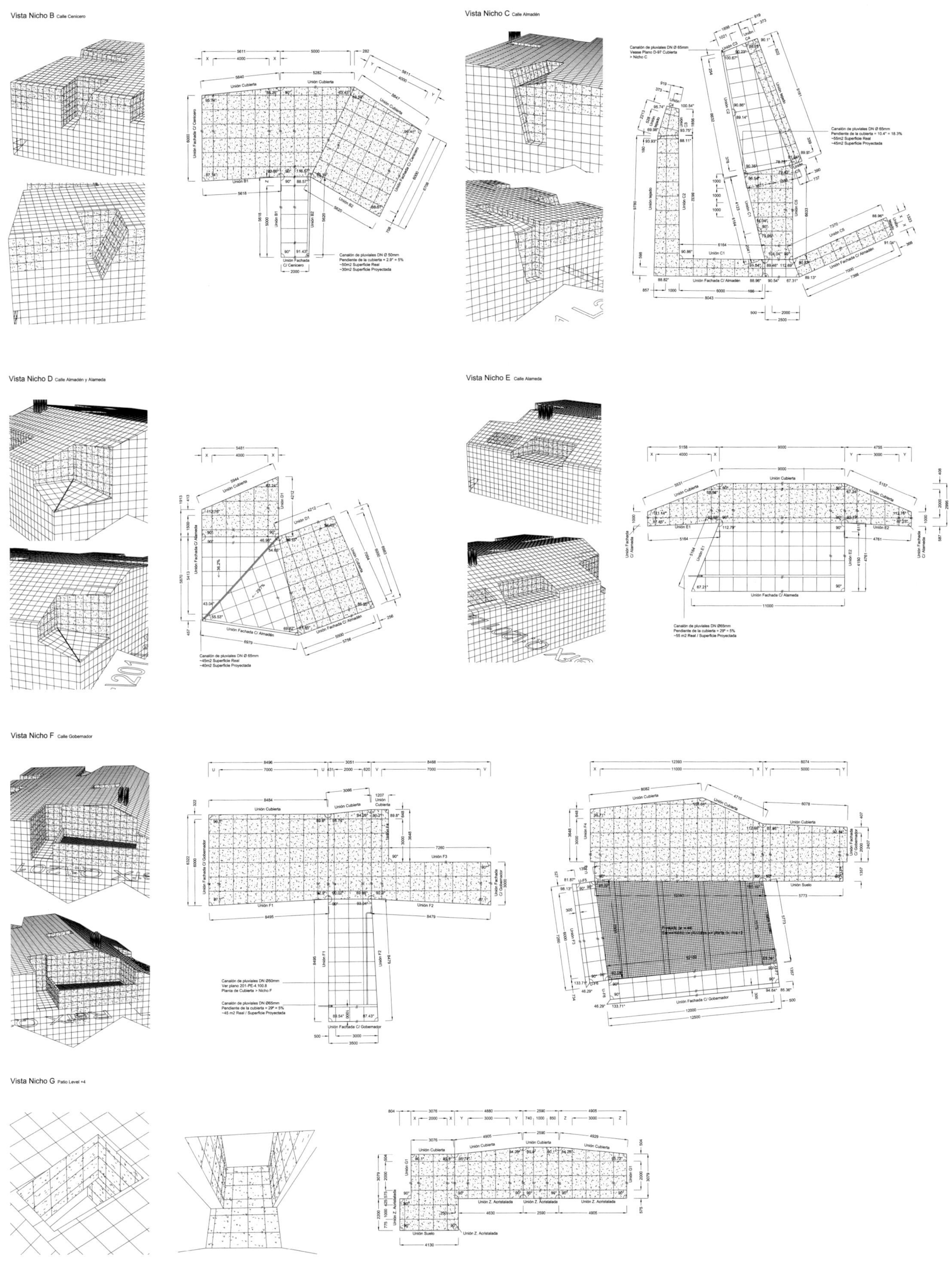

Vista Nicho B Calle Cenicero
Canalón de pluviales DN Ø 50mm
Pendiente de la cubierta = 2.8° = 5%
~50m2 Superficie Real
~30m2 Superficie Proyectada
Vista Nicho C Calle Almadén
Canalón de pluviales DN Ø 65mm
Pendiente de la cubierta = 10.4° = 18.3%
~55m2 Superficie Real
~45m2 Superficie Proyectada
Vista Nicho D Calle Almadén y Alameda
Canalón de pluviales DN Ø 65mm
~45m2 Superficie Real
~40m2 Superficie Proyectada
Vista Nicho E Calle Alameda
Canalón de pluviales DN Ø65mm
Pendiente de la cubierta = 29° = 5%
~55 m2 Real / Superficie Proyectada
Vista Nicho F Calle Gobernador
Canalón de pluviales DN Ø50mm
Canalón de pluviales DN Ø65mm
Pendiente de la cubierta = 29° = 5%
~45 m2 Real / Superficie Proyectada
Vista Nicho G Patio Level +4

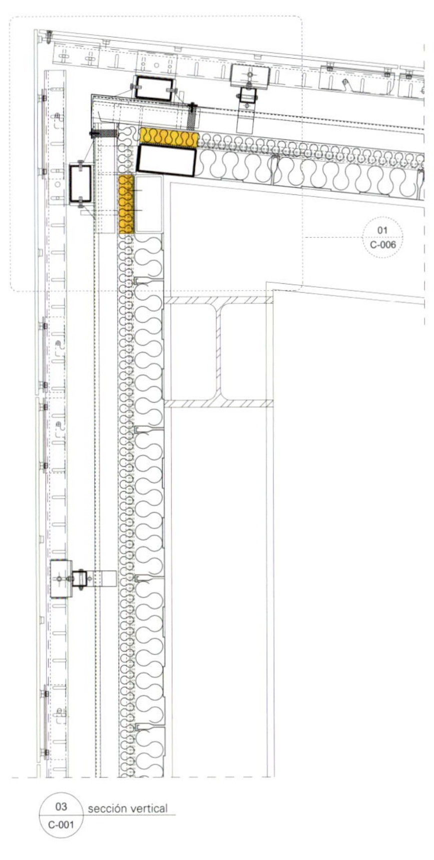

03 C-001 sección vertical

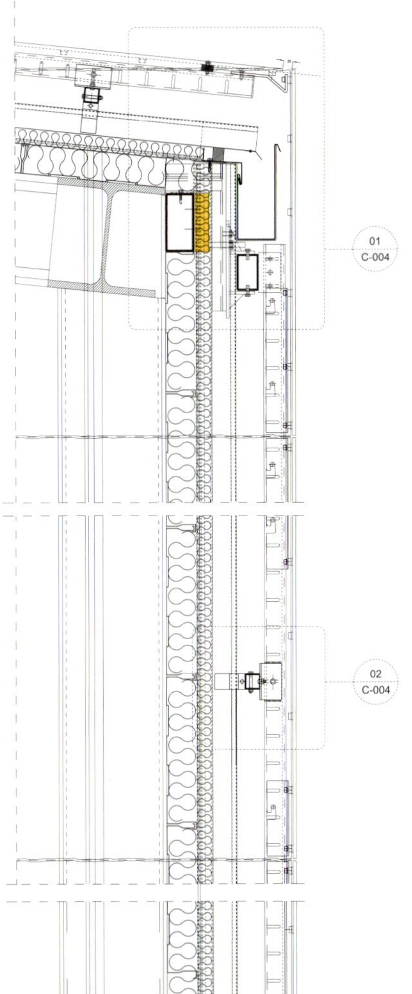

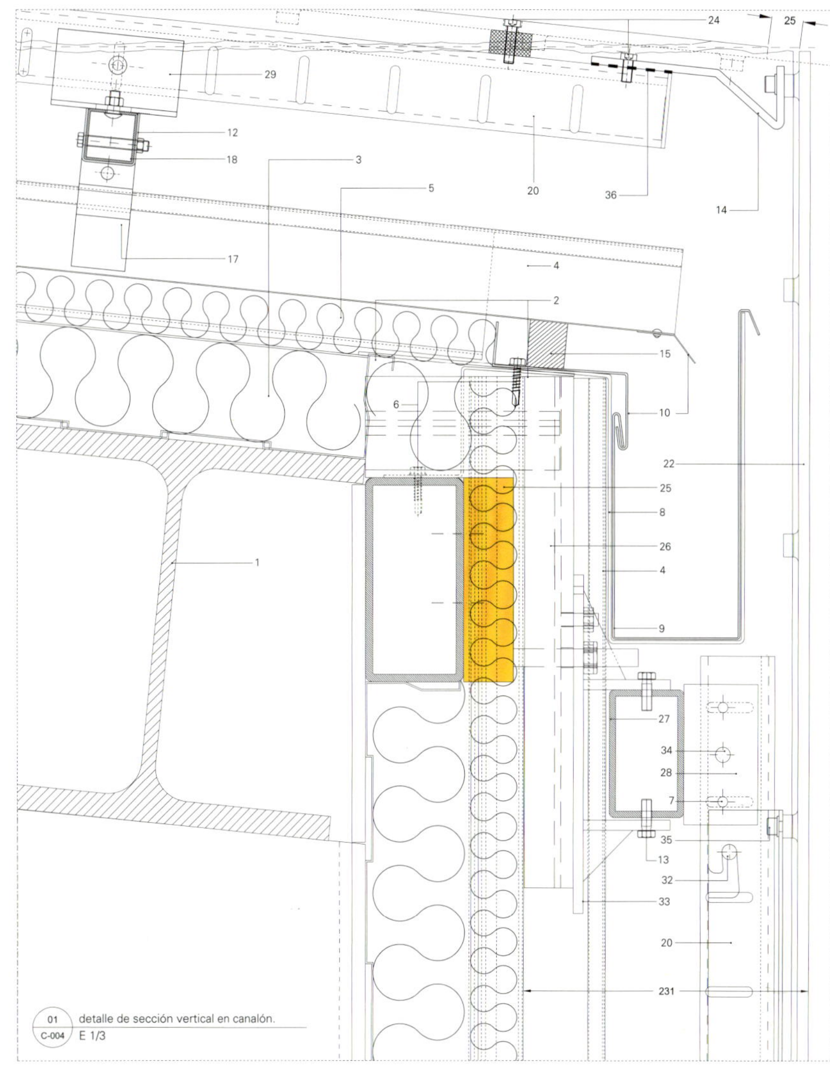

01 C-004 detalle de sección vertical en canalón.
E 1/3

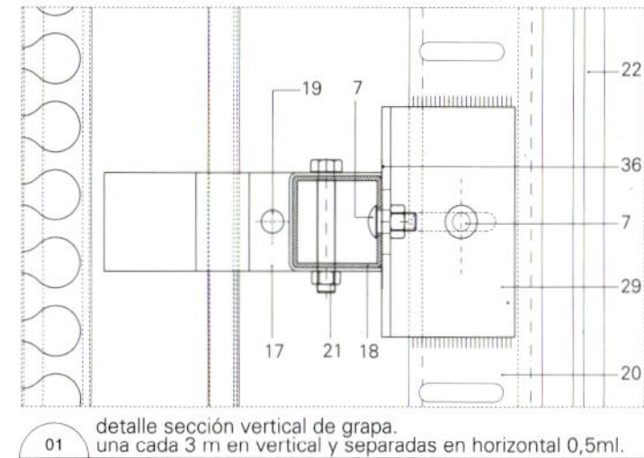

01 C-004 detalle sección vertical de grapa.
una cada 3 m en vertical y separadas en horizontal 0,5ml.
en cubierta se colocará cada 2 ml en longitudinal
y cada 0,5ml en transversal

1. estructura de hierro(no bellapart)
2. chapa de galvanizado
3. lana de roca de 80mm
4. chapa de inoxidable tipo bemo 0.8 lacada negra mate
5. lana de roca de 40mm
6. autotaladrante DIN 7504 D=6.3
7. tornillo DIN 603 M8x20+tuerca autobloc. Inox 304
8. fleje puntual fijación de canalón
9. canalón
10. chapa refuerzo de vierteaguas
11. tornillos DIN 933 M10 inoxAISI 304
12. neopreno
13. tornillo DIN 933 M8x25 Inox 304
14. chapa de 5mm inox-304
15. espuma expansiva tipo tecnobond, adhesiva a 1 cara
16. perfil de aluminio(gripi)
17. grapa de inoxidable de 3mm de grosor
18. perfil de inoxidable de 2mm en forma de U con colisos de ø8x15
19. tornillo allen DIN 912 M10x25 Inox 304
20. perfil en forma de U de 2.5mm
21. tornillo DIN 933 M8x60+tuerca autobloc. Inox
22. chapa de fundición
23. chapa de inoxidable AISI 304 de 0.8mm
24. allen DIN 7984 de inoxidable M8x40
25. macizo de hierro de 60x40
26. conjunto de U de 10mm de inox AISI 304+biga IPE 140 de hierro soldado a taller
27. tubo de inoxidable AISI 304 de 100x60x6
28. passamano de inox de 110x60x4mm
29. angulo de inoxidable AISI 304 de 4mm
30. arandela de EPDM
31. angulo de inox. en bruto de 3mm AISI 304
32. barilla roscada de M12 inox AISI 304
33. composición de chapas de inoxidable de 8mm con cartelas de refuerzo de 12mm AISI 304
34. tornillo DIN 933 y tuerca 934 inox AISI 304 M12
35. tornillo DIN 912 de M8x18 inox AISI 304
36. banda de tefión

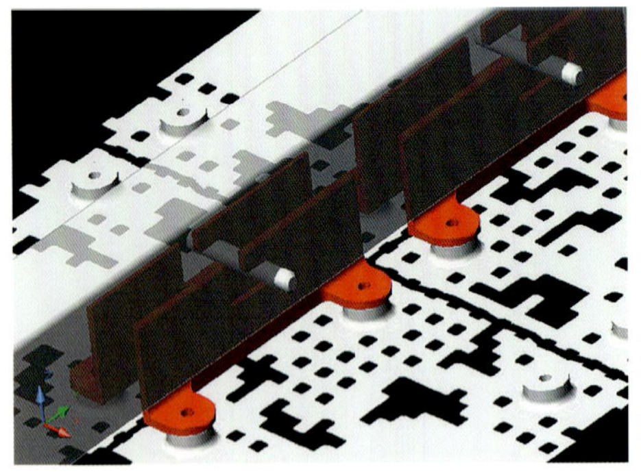

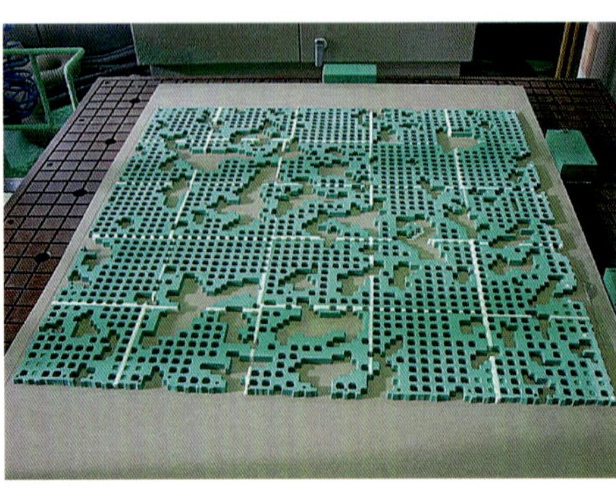

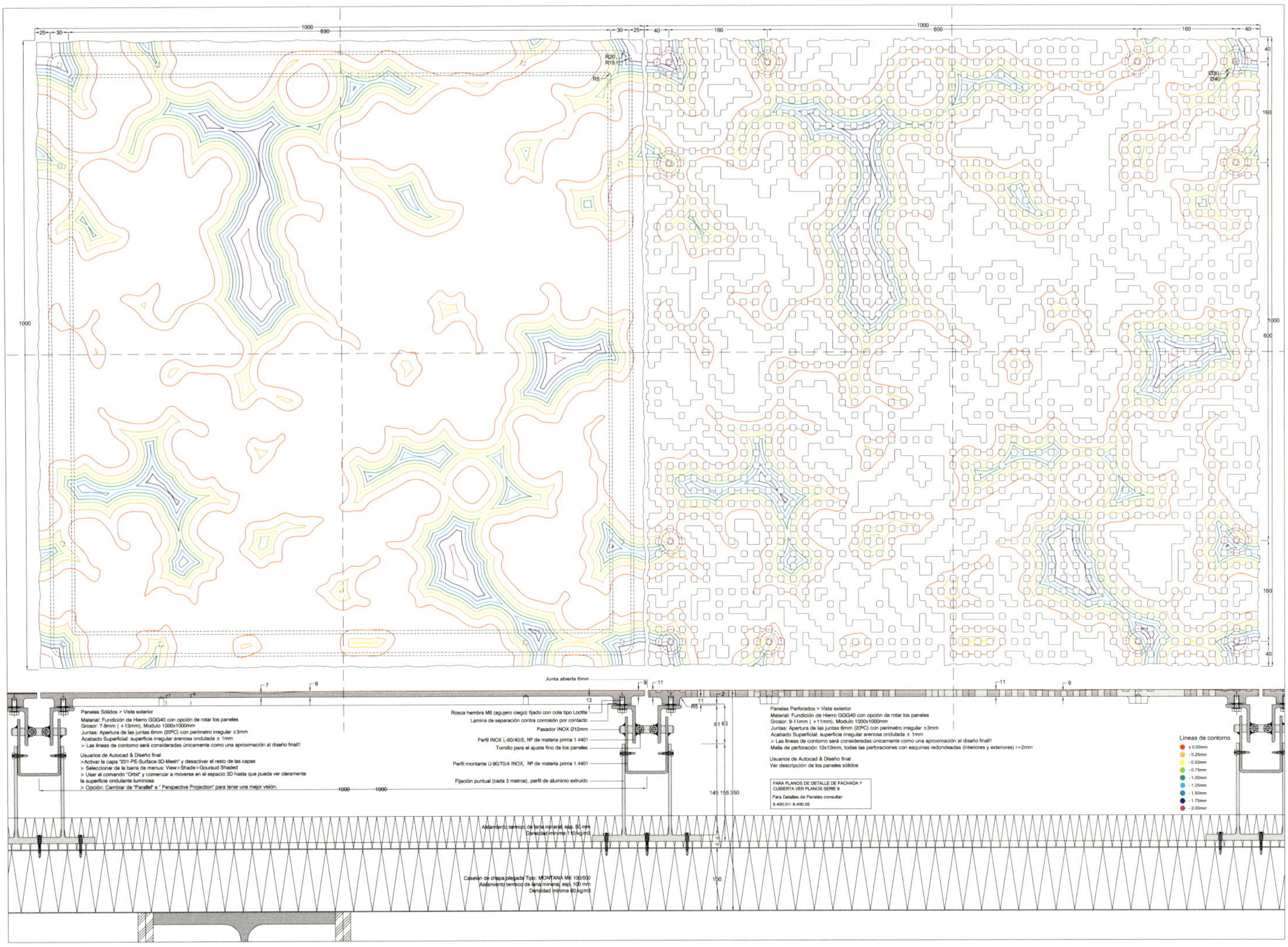

Junta abierta 6mm

Paneles Sólidos > Vista exterior
Material: Fundición de Hierro GGG40 con opción de rotar los paneles
Grosor: 7-9mm (+13mm), Modulo 1000x1000mm
Juntas: Apertura de las juntas 6mm (20ºC) con perímetro irregular ±3mm
Acabado Superficial: superficie irregular arenosa ondulada ± 1mm
> Las lineas de contorno será consideradas únicamente como una aproximación al diseño final!!

Usuarios de Autocad & Diseño final
>Activar la capa "201-PE-Surface-3D-Mesh" y desactivar el resto de las capas
> Seleccionar de la barra de menus: View>Shade>Gouraud Shaded
> Usar el comando "Orbit" y comenzar a moverse en el espacio 3D hasta que pueda ver claramente la superficie ondulante luminosa
> Opción: Cambiar de "Parallel" a " Perspective Projection" para tener una mejor visión.

Rosca hembra M8 (agujero ciego) fijado con cola tipo Loctite
Lamina de separación contra corrosión por contacto
Pasador INOX Ø12mm
Perfil INOX L-60/40/5, Nº de materia prima 1.4401
Tornillo para el ajuste fino de los paneles
Perfil montante U-90/70/4 INOX, Nº de materia prima 1.4401
Fijación puntual (cada 3 metros), perfil de aluminio extruido

Aislamiento termico de lana mineral, esp. 50 mm
Densidad minima 110 kg/m3

Casetón de chapa plegada Tipo: MONTANA MK 100/500
Aislamiento termico de lana mineral, esp. 100 mm
Densidad minima 60 kg/m3

Paneles Perforados > Vista exterior
Material: Fundición de Hierro GGG40 con opción de rotar los paneles
Grosor: 9-11mm (+11mm), Modulo 1000x1000mm
Juntas: Apertura de las juntas 6mm (20ºC) con perímetro irregular ±3mm
Acabado Superficial: superficie irregular arenosa ondulada ± 1mm
> Las lineas de contorno será consideradas únicamente como una aproximación al diseño final!!
Malla de perforación 10x10mm, todas las perforaciones con esquinas redondeadas (interiores y exteriores) r=2mm

Usuarios de Autocad & Diseño final
Ver descripción de los paneles sólidos

PARA PLANOS DE DETALLE DE FACHADA Y CUBIERTA VER PLANOS SERIE 9.
Para Detalles de Paneles consultar:
9.400.01/ 9.400.02

Lineas de contorno
± 0.00mm
- 0.25mm
- 0.50mm
- 0.75mm
- 1.00mm
- 1.25mm
- 1.50mm
- 1.75mm
- 2.00mm

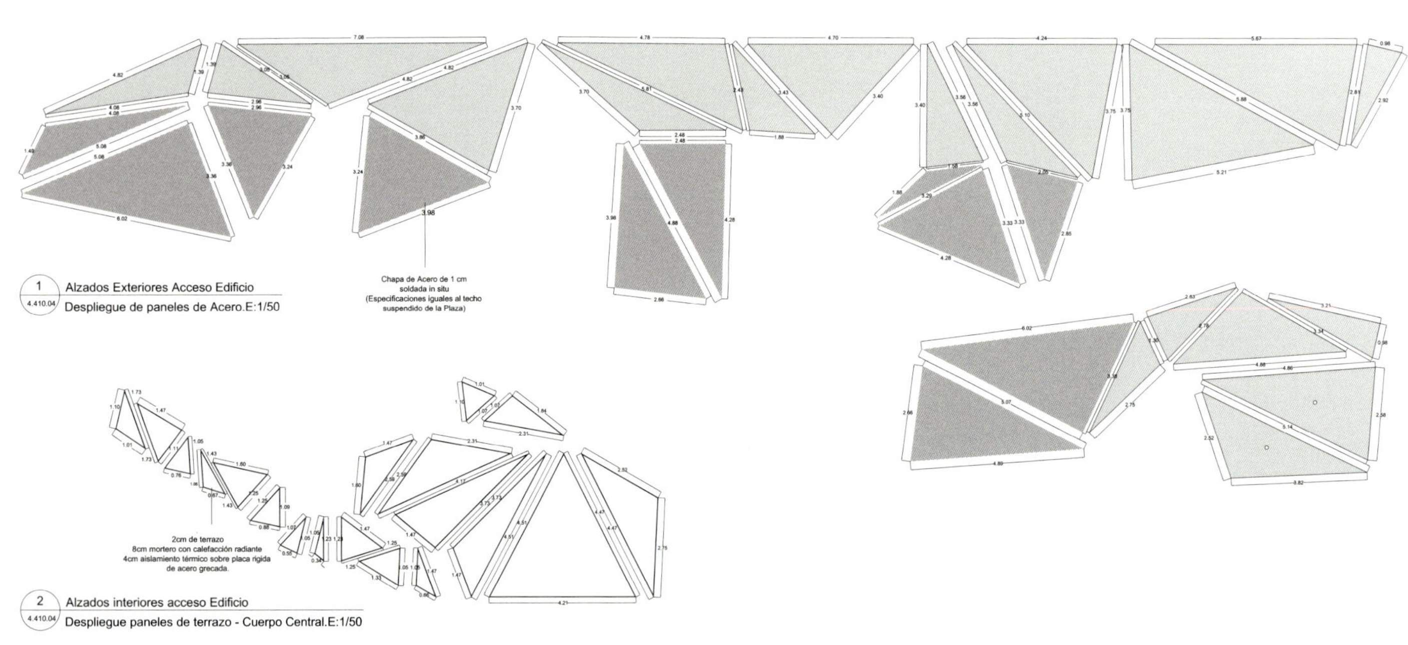
1
4.410.04
Alzados Exteriores Acceso Edificio
Despliegue de paneles de Acero.E:1/50
Chapa de Acero de 1 cm
soldada in situ
(Especificaciones iguales al techo
suspendido de la Plaza)
2cm de terrazo
8cm mortero con calefacción radiante
4cm aislamiento térmico sobre placa rigida
de acero grecada.
2
4.410.04
Alzados interiores acceso Edificio
Despliegue paneles de terrazo - Cuerpo Central.E:1/50

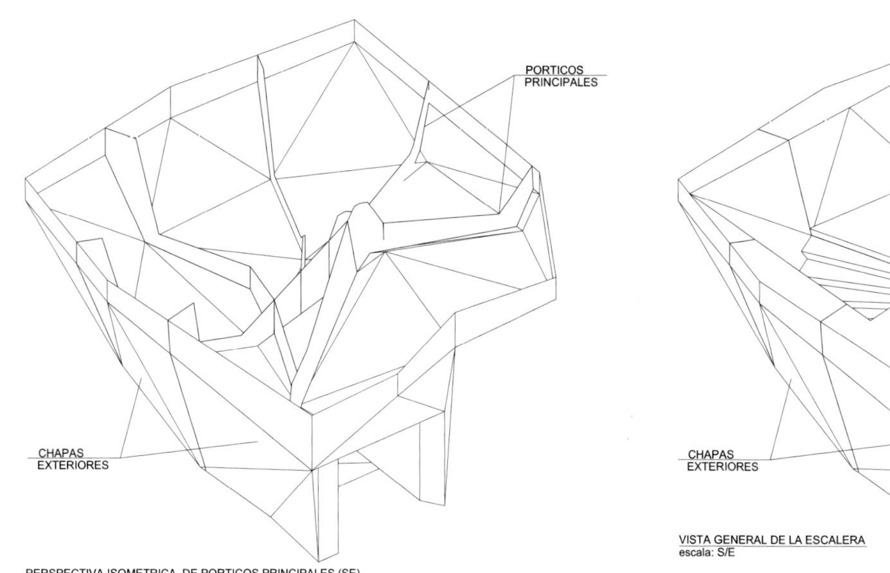

PERSPECTIVA ISOMETRICA DE PORTICOS PRINCIPALES (SE)

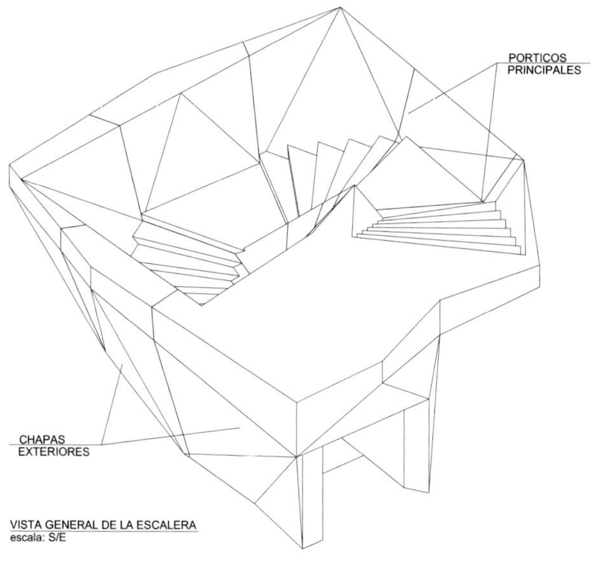

VISTA GENERAL DE LA ESCALERA
escala: S/E

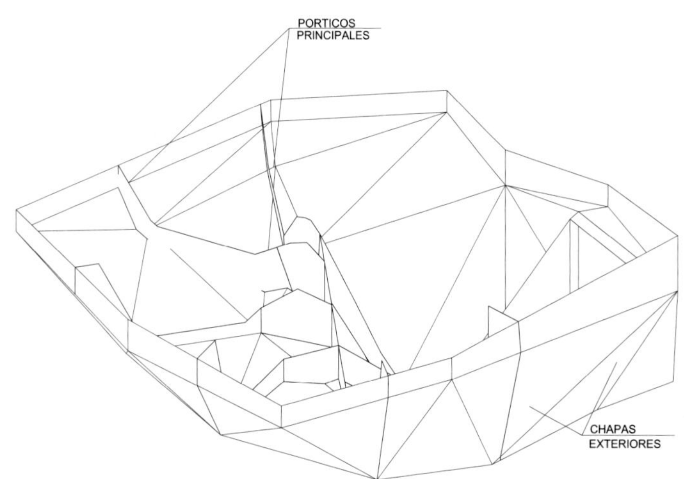

PERSPECTIVA ISOMETRICA DE PORTICOS PRINCIPALES (SO)

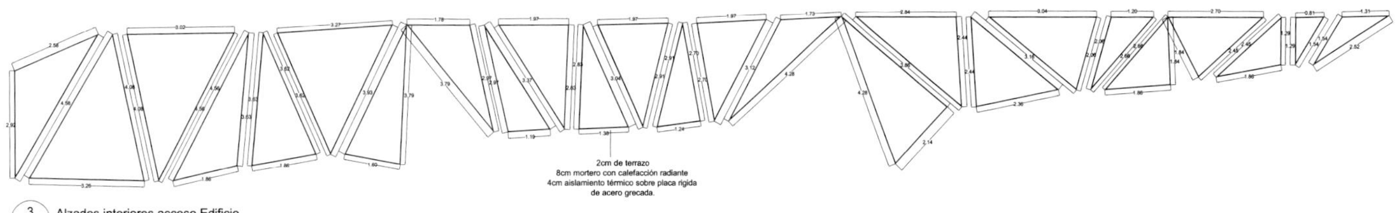

3 Alzados interiores acceso Edificio
4.410.04 Despliegue paneles de terrazo - Cuerpo Exterior.E:1/50

2 cm de terrazo in situ
8 cm mortero con calefacción radiante
4 cm de aislamiento térmico sobre placa grecada rígida de acero
1.5 mm perfil de aluminio
1.5 cm placa de Acero soldada in situ
Estructura primaria formada por placa de acero de2 cm soldada en obra.

4 Escalera de Acceso al Edificio 1:10
4.410.04 Planta detalle de cerramiento interior.

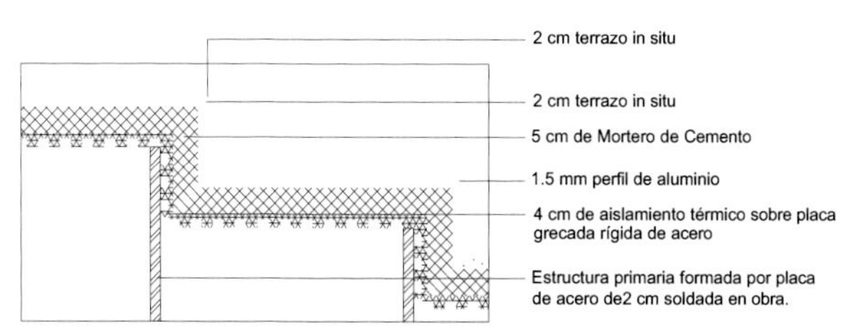

5 Escalera de Acceso al Edificio 1:10
4.410.04 Sección Constructiva del Peldañeado

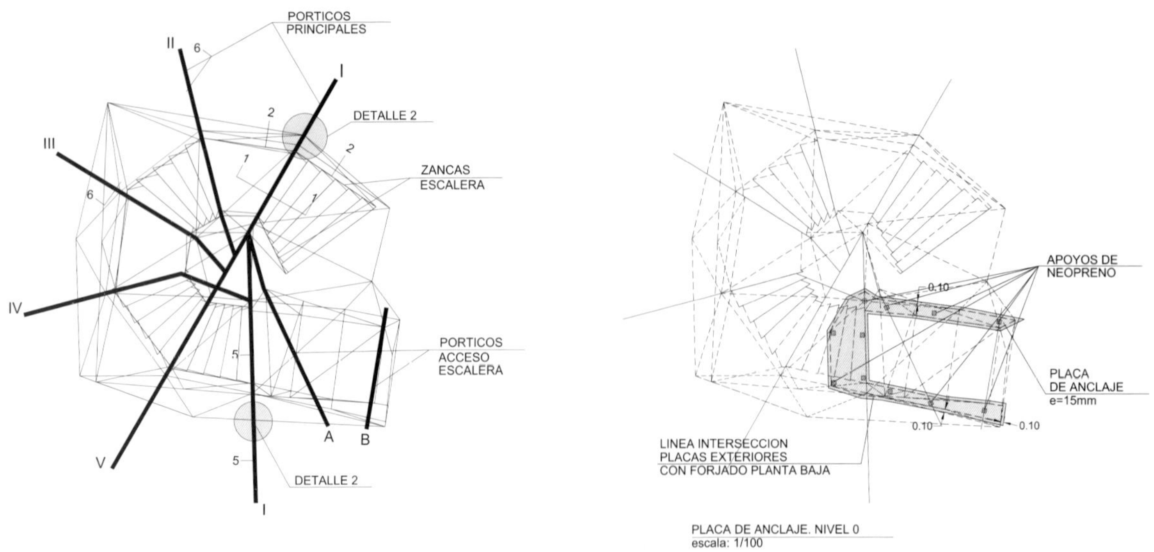
PORTICOS
PRINCIPALES
DETALLE 2
ZANCAS
ESCALERA
PORTICOS
ACCESO
ESCALERA
DETALLE 2
APOYOS DE
NEOPRENO
PLACA
DE ANCLAJE
e=15mm
LINEA INTERSECCION
PLACAS EXTERIORES
CON FORJADO PLANTA BAJA
PLACA DE ANCLAJE. NIVEL 0
escala: 1/100

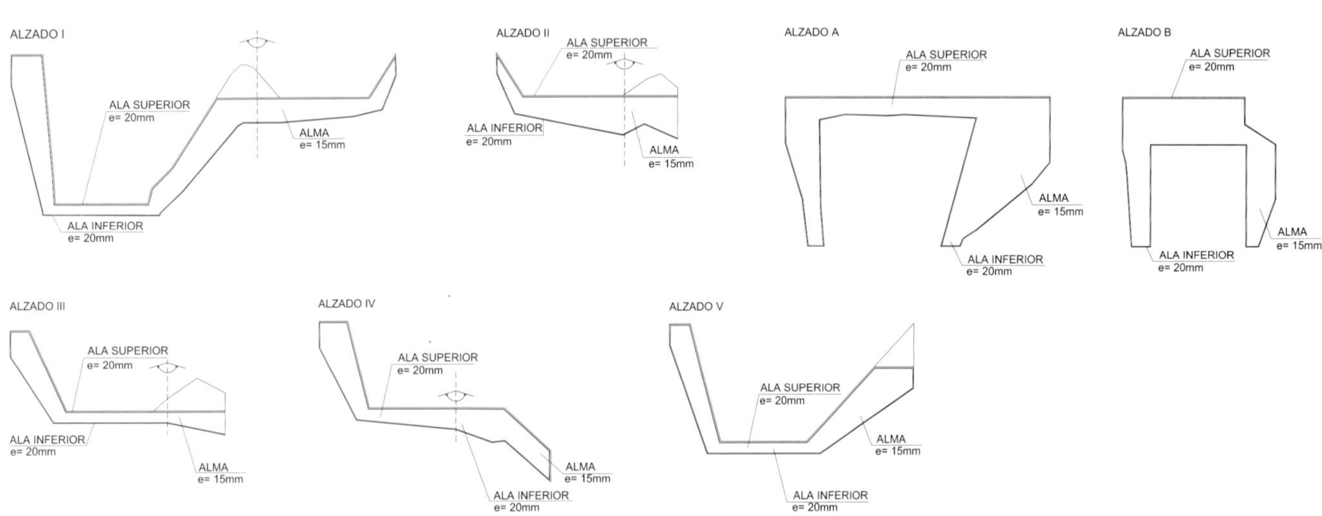
ALZADO I
ALA SUPERIOR
e= 20mm
ALMA
e= 15mm
ALA INFERIOR
e= 20mm
ALZADO II
ALZADO A
ALZADO B
ALZADO III
ALZADO IV
ALZADO V

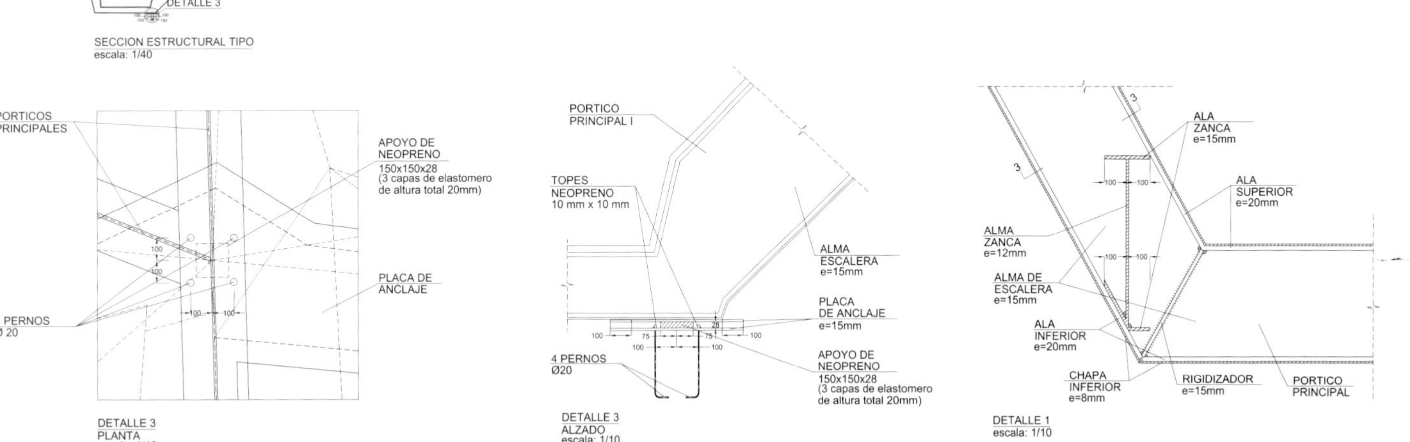
DETALLE 2
DETALLE 1
DETALLE 3
SECCION ESTRUCTURAL TIPO
escala: 1/40
PORTICOS
PRINCIPALES
APOYO DE
NEOPRENO
150x150x28
(3 capas de elastomero
de altura total 20mm)
PLACA DE
ANCLAJE
4 PERNOS
Ø 20
DETALLE 3
PLANTA
escala: 1/10
PORTICO
PRINCIPAL I
TOPES
NEOPRENO
10 mm x 10 mm
ALMA
ESCALERA
e=15mm
PLACA
DE ANCLAJE
e=15mm
4 PERNOS
Ø20
DETALLE 3
ALZADO
escala: 1/10
ALA
ZANCA
e=15mm
ALA
SUPERIOR
e=20mm
ALMA
ZANCA
e=12mm
ALMA DE
ESCALERA
e=15mm
ALA
INFERIOR
e=20mm
CHAPA
INFERIOR
e=8mm
RIGIDIZADOR
e=15mm
PORTICO
PRINCIPAL
DETALLE 1
escala: 1/10

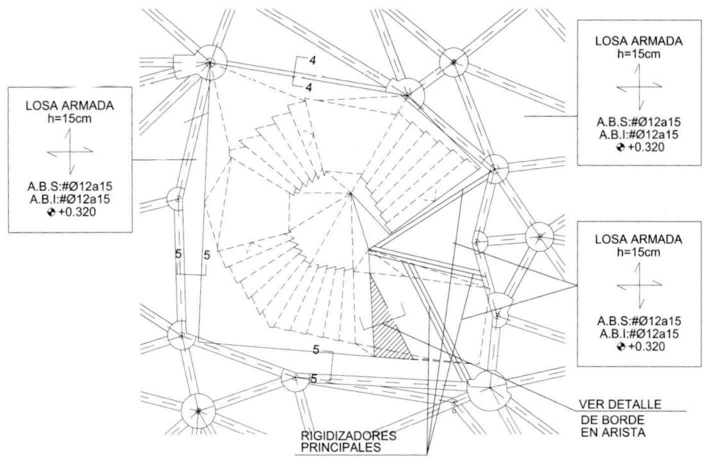

FORJADO PLANTA PRIMERA
escala: 1/10

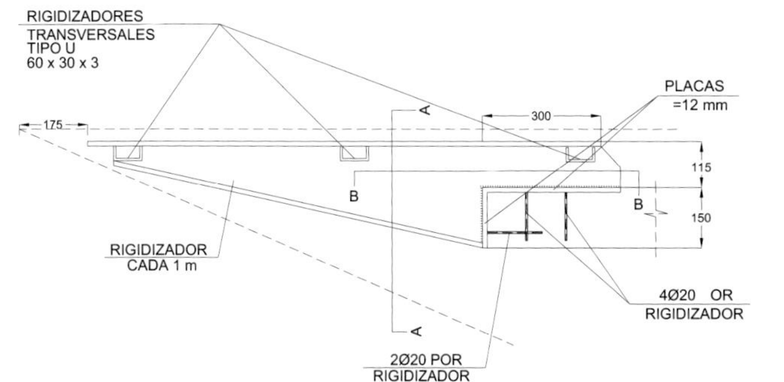

DETALLE DE BORDE EN ARISTA
escala: 1/10

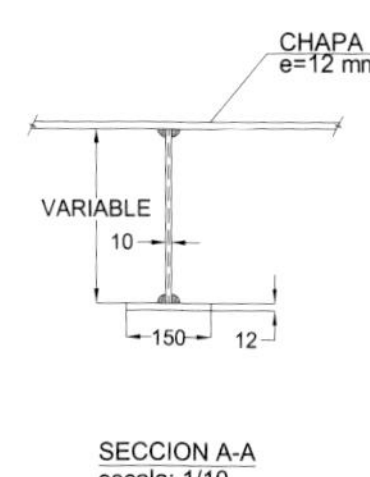

SECCION A-A
escala: 1/10

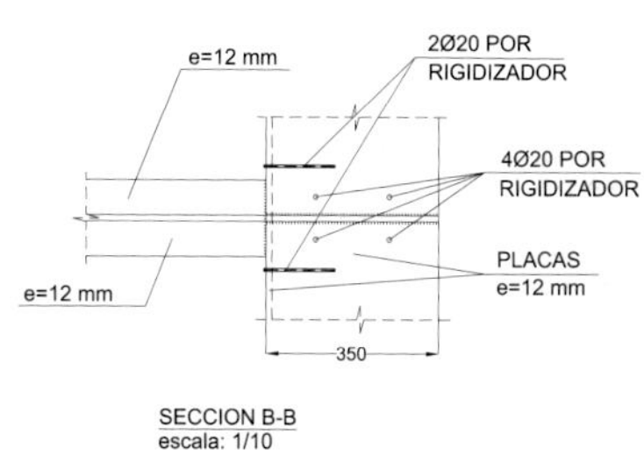

SECCION B-B
escala: 1/10

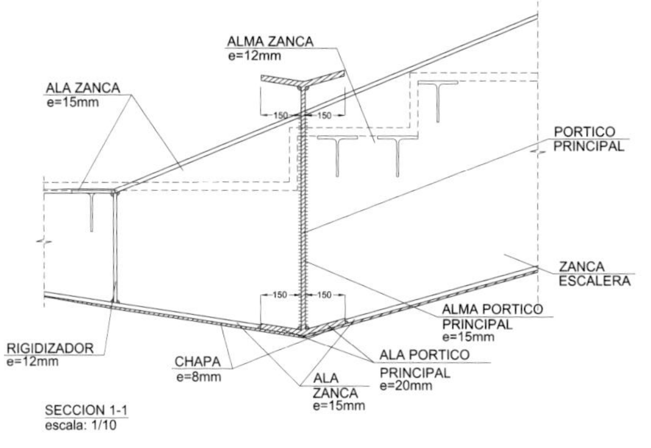

SECCION 1-1
escala: 1/10

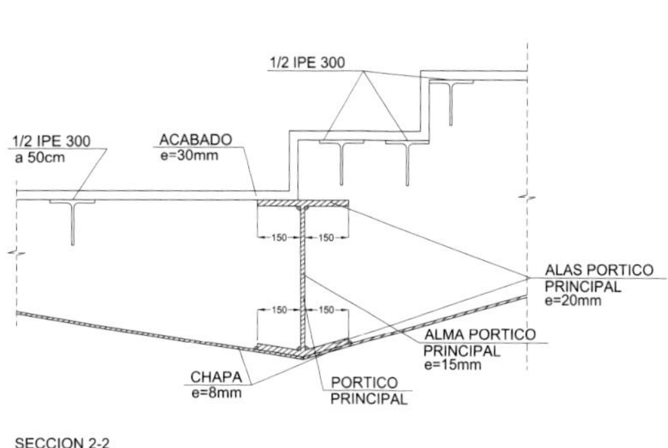

SECCION 2-2
escala: 1/10

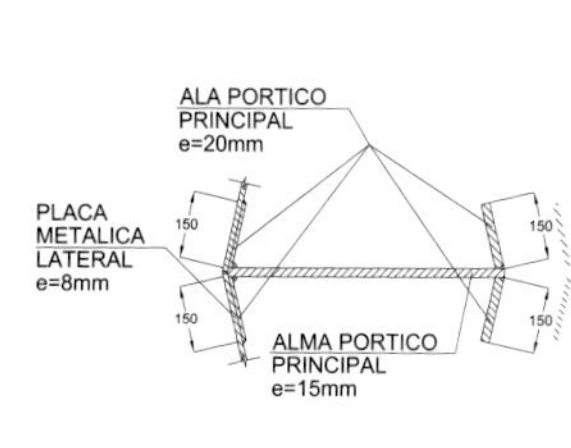

SECCION 3-3
PORTICO PRINCIPAL (Planta)
escala: 1/10

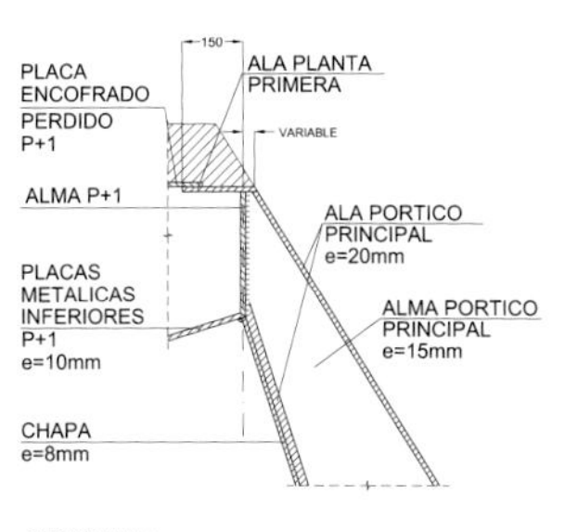

SECCION 4-4
DETALLE 2
ENCUENTRO CON PLANTA +1
escala: 1/10

PLACAS ENCOFRADO
PERDIDO P+1

ALA PLANTA
PRIMERA

ALMA
PLANTA
PRIMERA

ALA PORTICO
PRINCIPAL
e=20mm

PLACA
METALICA
INFERIORES
P+1
e=10mm

ALMA PORTICO
PRINCIPAL
e=15mm

CHAPA
e=8mm

SECCION 5-5
DETALLE 2
ENCUENTRO CON PLANTA +1
escala: 1/10

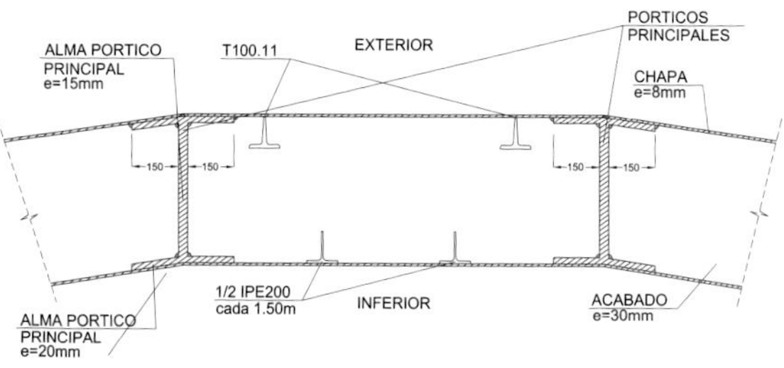

SECCION 6-6
DETALLE SUJECCION DE ACABADOS Y CHAPA EXTERIOR
escala: 1/10

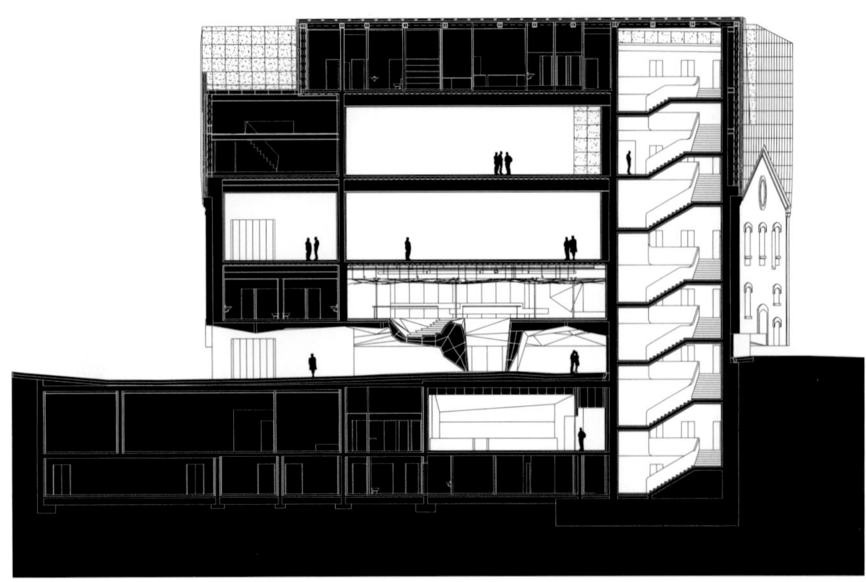

section A - A'

0 2 5m

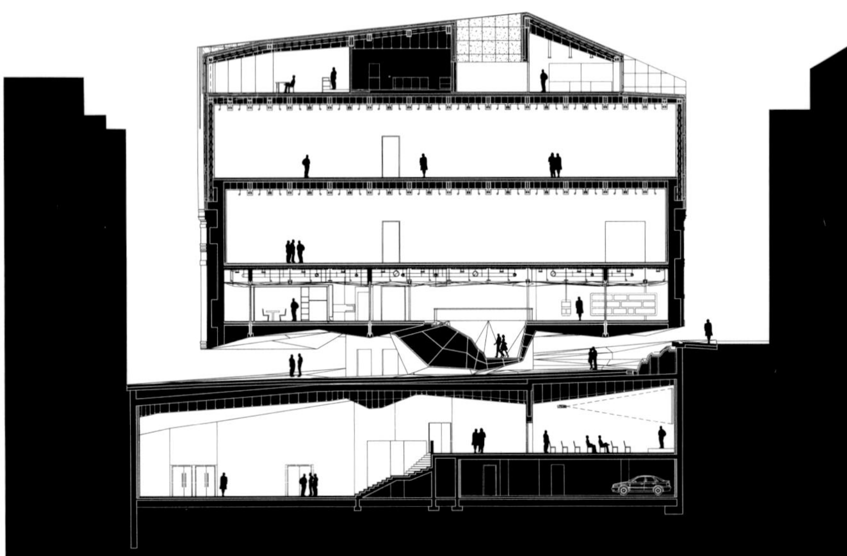

section B - B'

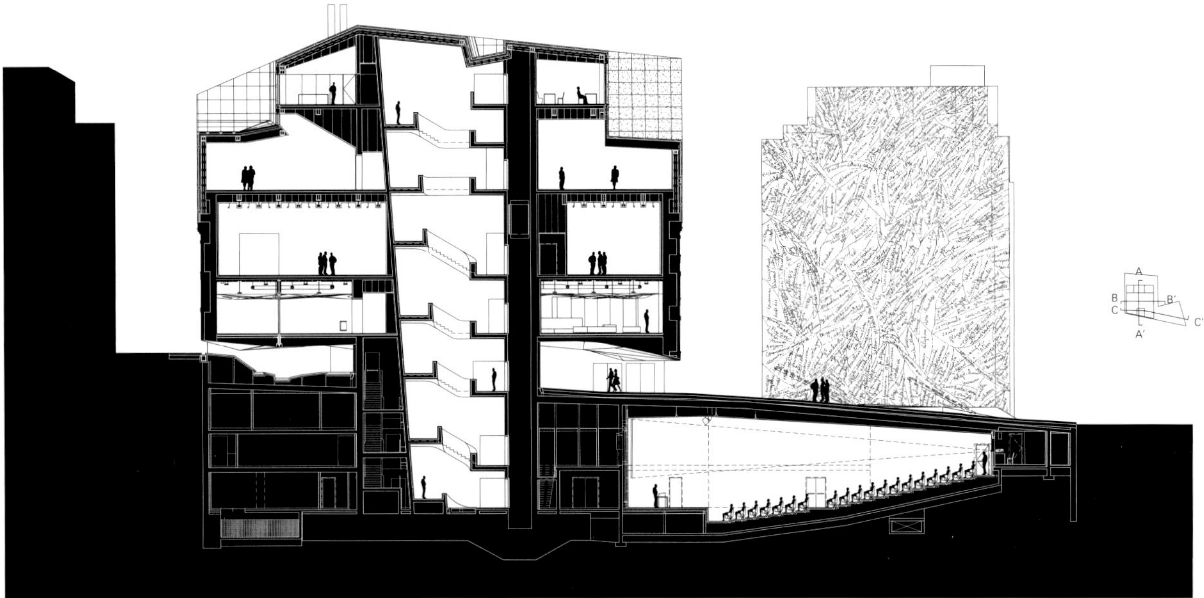

section C - C'

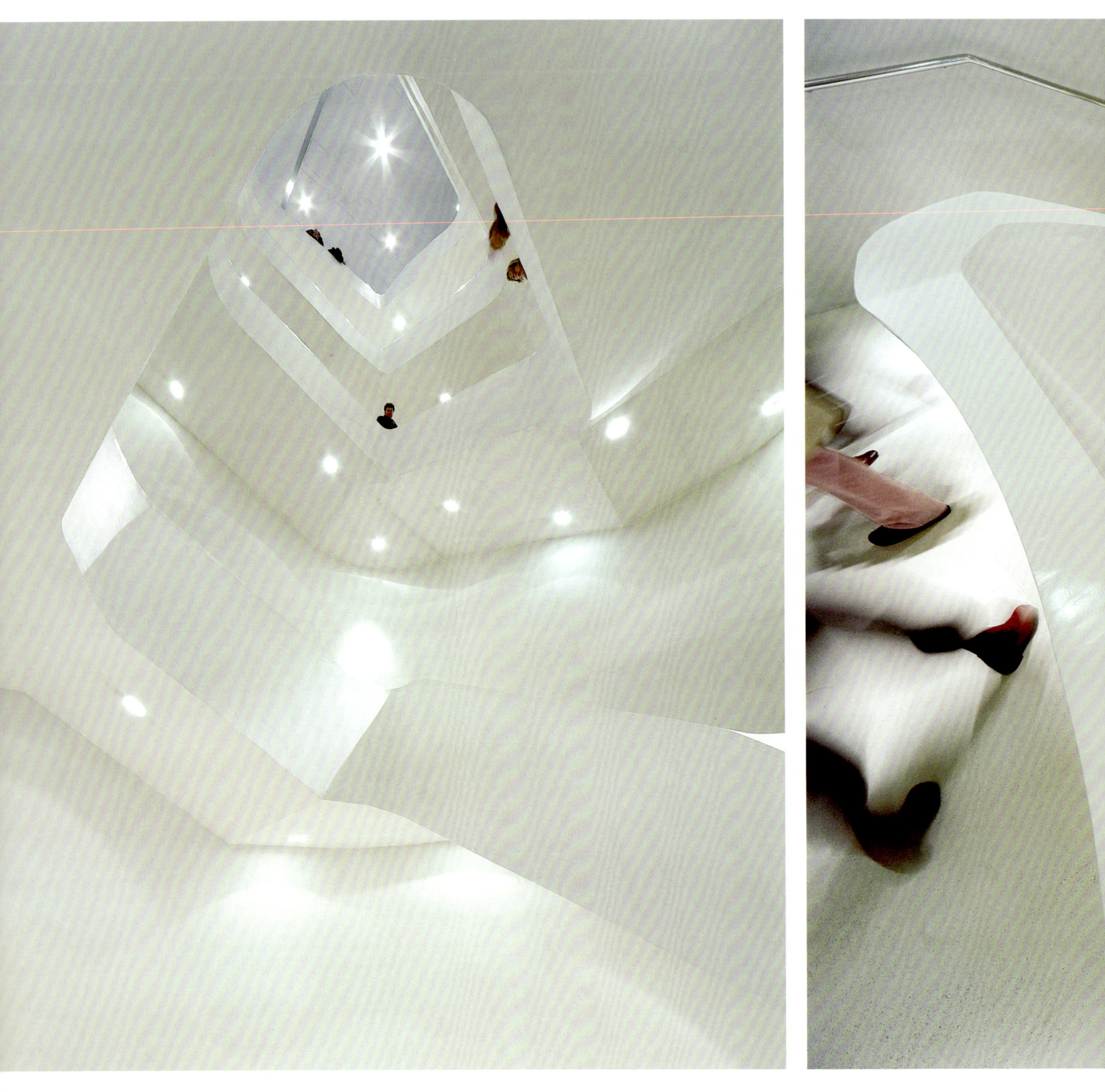

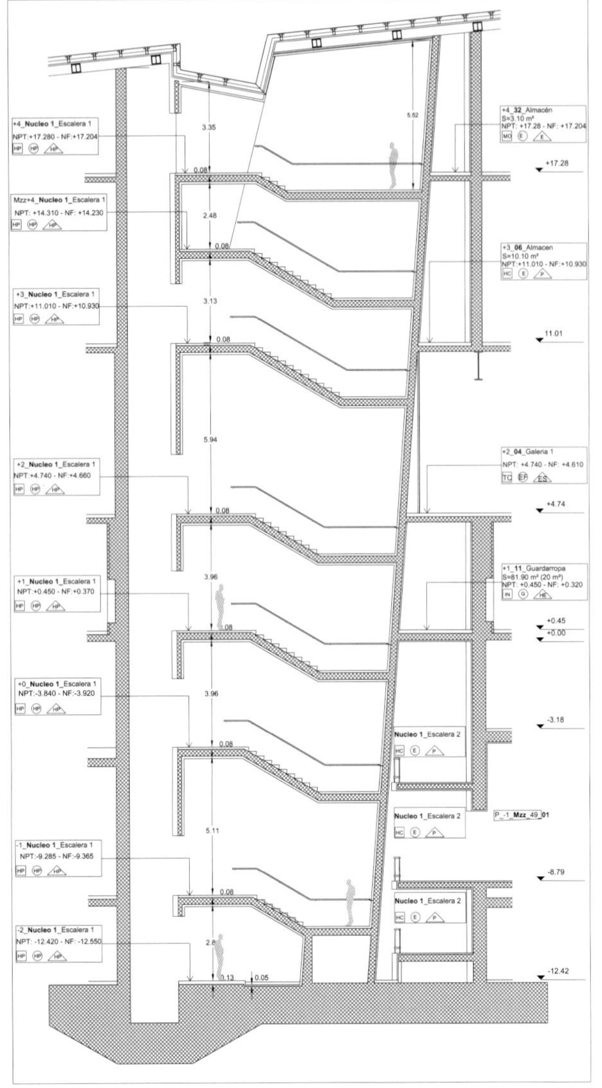

seccion parcial A - A'

EL HORMIGÓN SE ADAPTARÁ A LA POSICIÓN DE LA FABRICA DE LADRILLO EXISTENTE.

seccion parcial B - B'

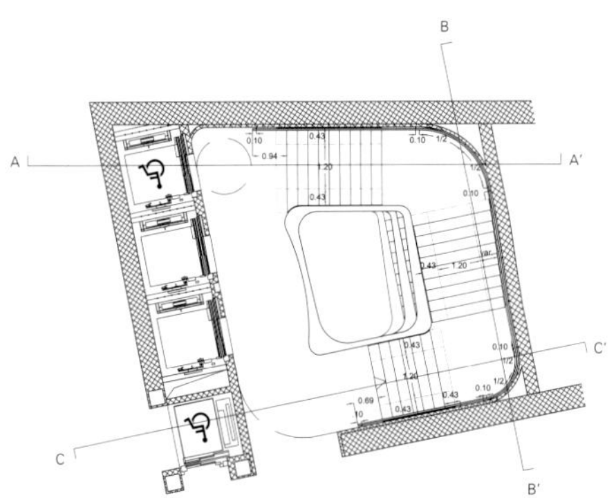

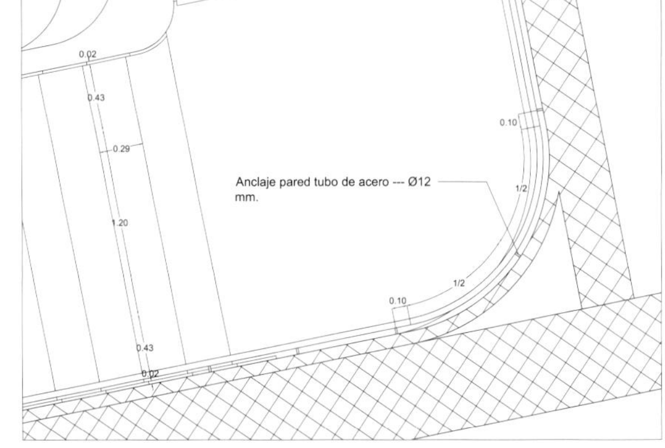

planta tipo. replanteo barandilla curva y peldaño

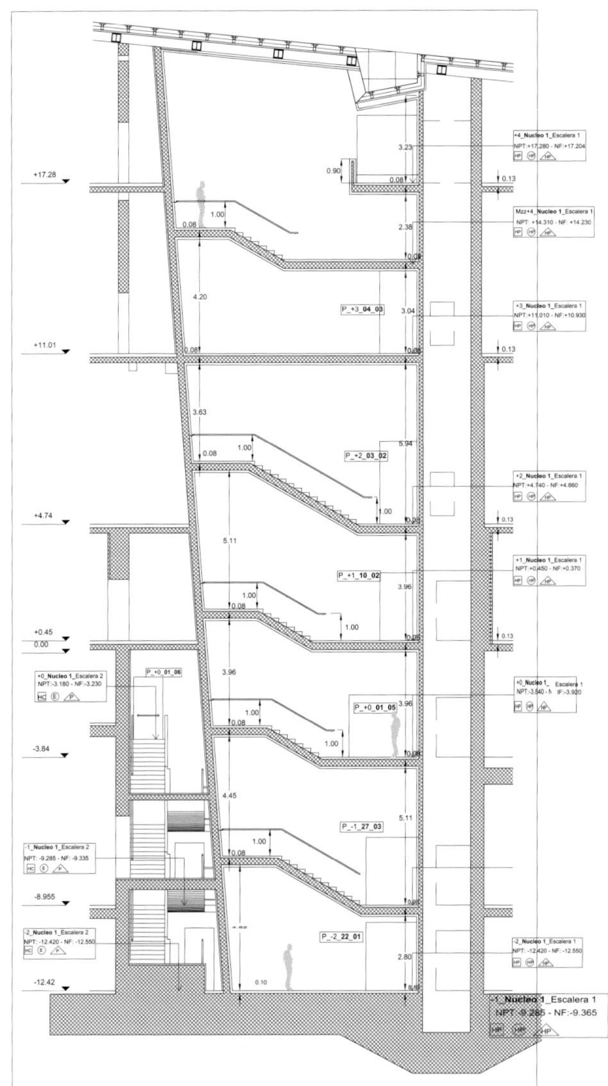
seccion parcial C - C'

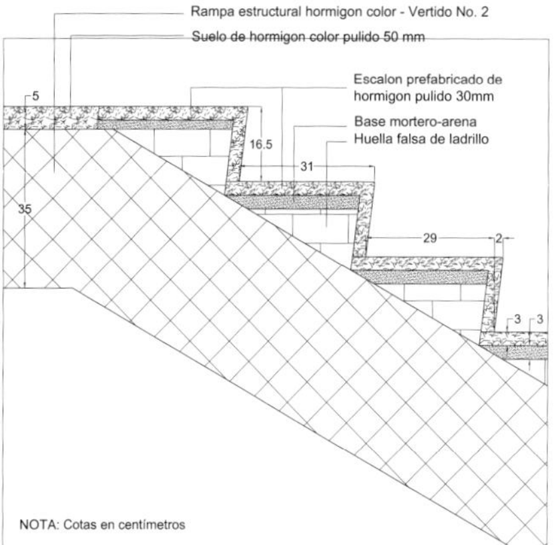

rampa - escalon

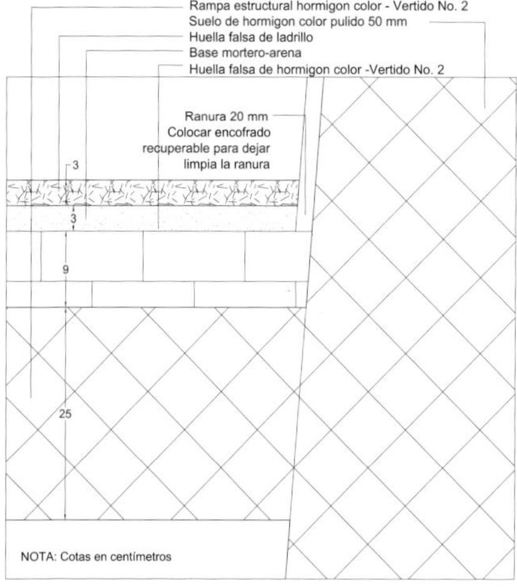

escalon - ranura

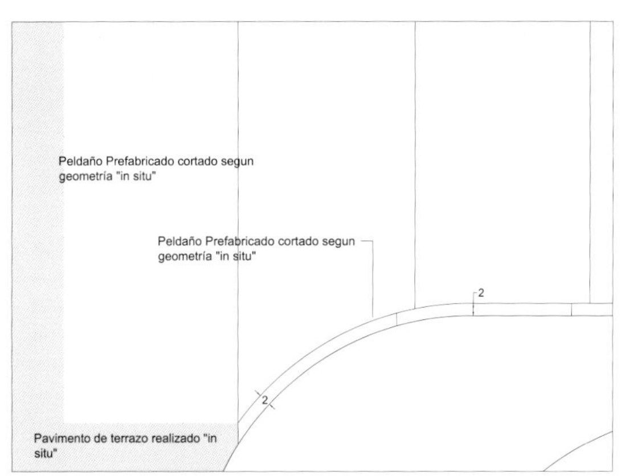

planta tipo. replanteo barandilla y peldaño

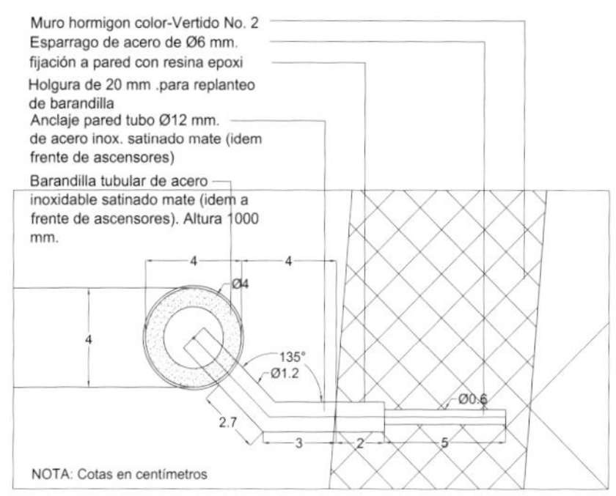

barandilla

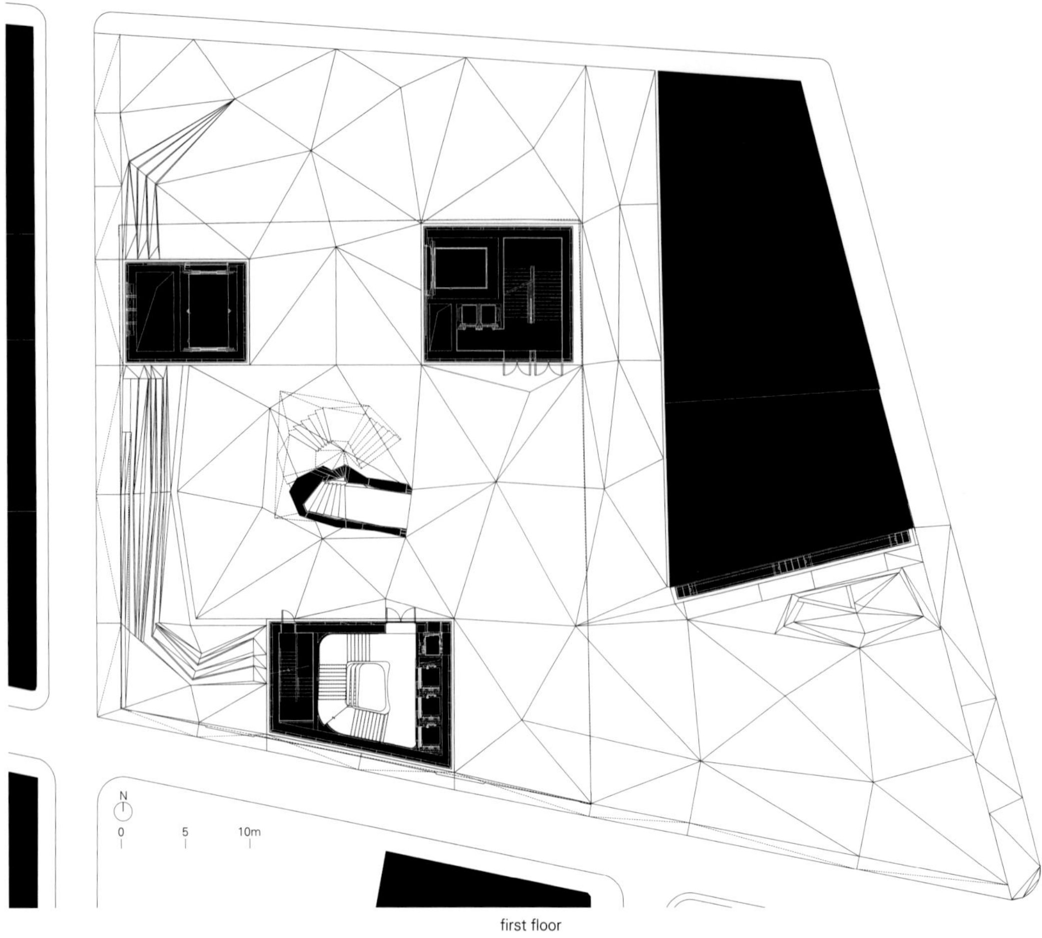

first floor

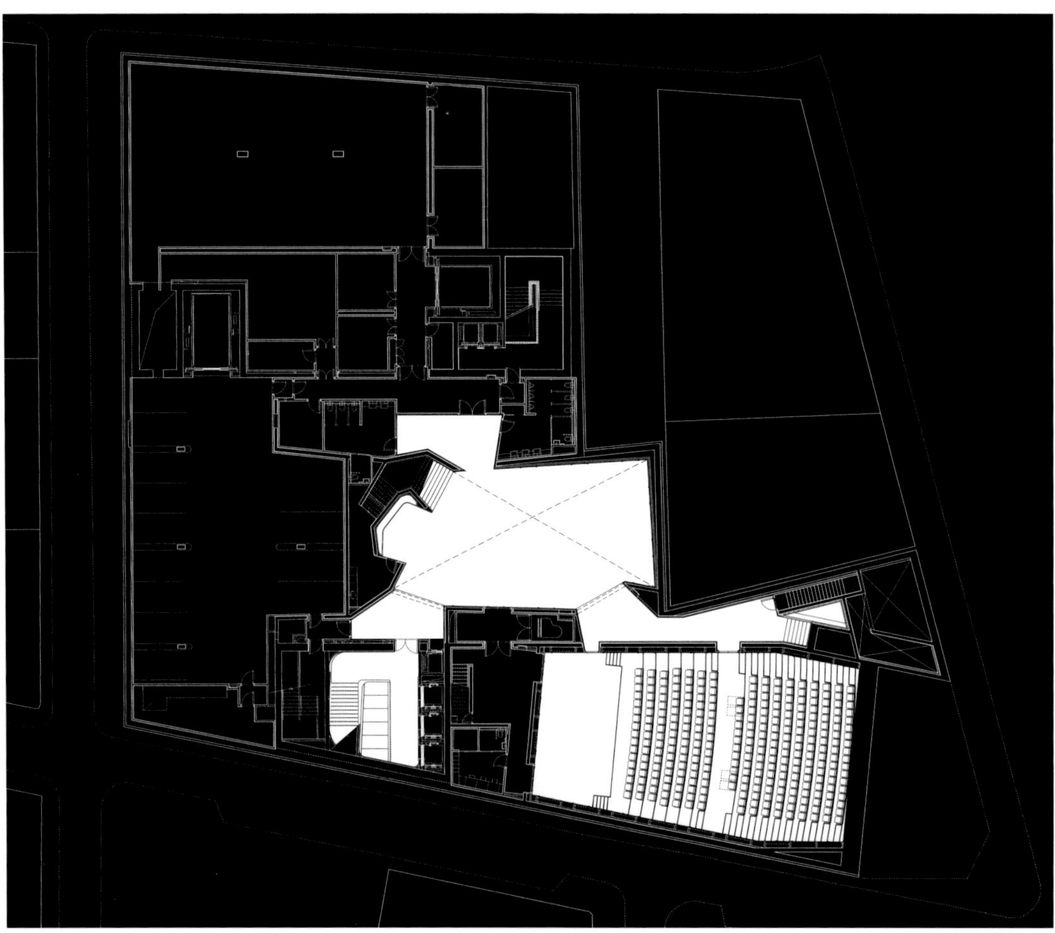

second floor below ground

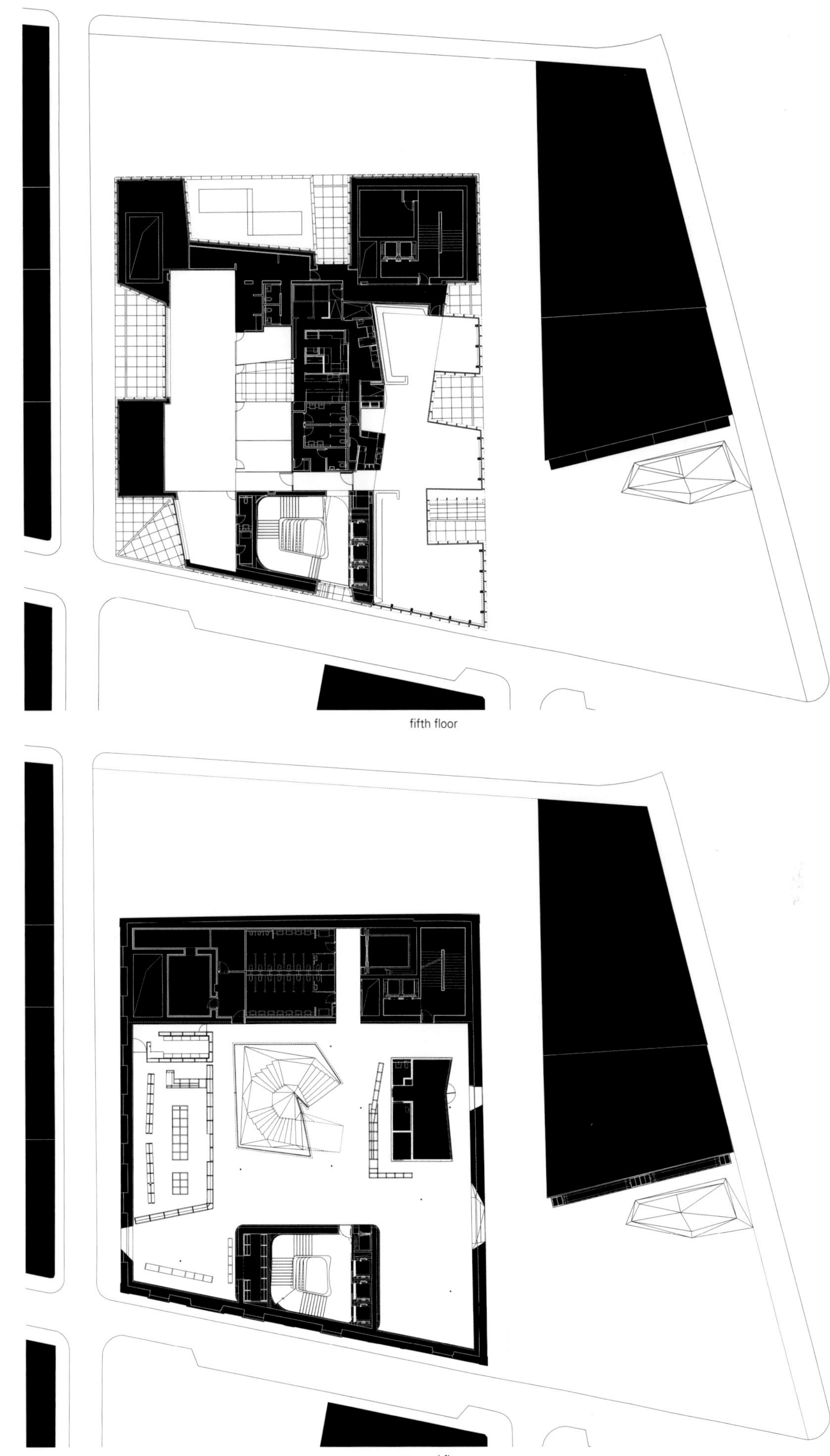

fifth floor

second floor

On the premises of the Westfalian company glaskoch who distributes glass products under the name "Leonardo" a significant corporate architecture was created that now forms a central element in the brand's overall communicative presence. As an atmospheric brandworld, the Leonardo Glass Cube conveys to guests and the staff alike the company's philosophy and visions in an inspiring manner. The open floor plan layout of the clearly designed and multi-functional Leonardo building enables an integrative linkage of product presentation zones, seminar and meeting rooms, inspiring work areas and a lot more besides across a total area of 1,200 square meters.

The glass facade of the building represents the passage to a hyper-naturalistic world with heightened aesthetic appeal. A transparent print slides into the insight or outlook as a subtly visible image plane. The graphically illustrated elements displayed on it were derived from the architecture and the surrounding landscape. They create a subtle puzzle, mingling with the reflections of their models in reality.

The edificial structure consists of two formally contrasting elements: A geometrically stringent, cube-like shell volume and a freeform positioned centrally in the interior. The undulating, curved white wall encases an introverted exhibition space and its other side circumscribes the

레오나르도 글래스 큐브

베스트팔리안의 유리 식기 회사인 레오나르도 사는 브랜드의 종합적인 정보를 전달하는 핵심 요소로 그 이름을 딴 사옥을 지었다. 레오나드로 글래스 큐브는 브랜드의 특성에 따라 방문객과 직원 모두에게 영감을 주며 회사의 철학과 비전을 전달한다. 하나의 열린 공간에 다목적으로 설계된 이 건물은 바닥 면적이 총 1,200 m²로 상품 전시 공간, 세미나 및 회의실, 작업실, 그 밖의 여러 공간을 포함하고 있다.

건물의 유리 입면은 한층 강조된 예술적 표현을 통해 초자연적 세계로 안내한다. 보일 듯 말 듯한 이미지를 담은 투명한 프린트가 입면 내외부에 스며들어 있다. 그래픽 디자인 요소들은 건물과 그 주변 경관에서 이끌어낸 것이다. 이들은 반사된 실제 영상과 혼재되어 미묘한 혼동을 준다.

건물의 구조는 대조를 이루는 두 가지 요소로 이루어져 있다. 하나는 엄중한 기하학의 정육면체 외피이며, 다른 하나는 자유롭게 배치된 내부공간이다. 흰 벽은 굽이치는 곡선을 그리며 내향적인 전시 공간을 감싸고, 그 반대편으로는 유리 입

3deluxe

Leonardo Glass Cube

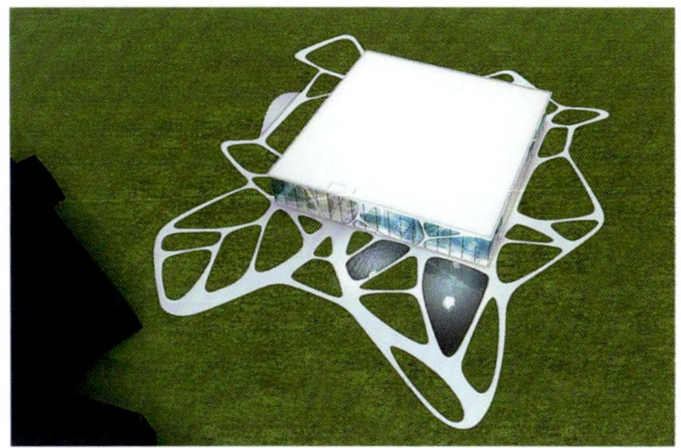

extroverted hallway along the glass facade. Three white sculptural structures - so-called 'Genetics' - connect the separate zones of the building to each other again.
On the glass facade 'Genetics' appear again in a two-dimensional version. The superimposed pilaster strips are continued in a network of white concrete pathways that surrounds the entire building and lets it grow together with its location.
In the centre of the interior ground floor and basement are connected by a void crossed by bridges. Entering the Glass Cube through the ground-floor main entrance, visitors encounter a space that opens up not just horizontally, but also upwards and downwards. On both floors the wall rolls in to form niches that are used for various functions such as themed product orchestrations and meeting lounges. In particular in the breaks in the wall these lines predominate as a significant graphic design element that is continued on the ceiling as a system of ventilation joints. On the side facing the facade, the material nature of the white surface is visually dissolved by means of a layer of gauze suspended in front. Dynamically programmed artificial light as well as the incidence of daylight sets colour highlights in the purely white interior and create a permanent change of atmosphere.

면을 따라 외향적인 복도를 형성한다. 소위 '유전자 구조'라고 불리는 세 개의 흰색 조형 구조물은 분리되어 있는 공간들을 서로 연결한다.
이 '유전자 구조'는 유리 입면 위에서 2차원으로 한번 더 나타난다. 외벽에 덧붙인 가느다란 벽기둥은 건물 전체를 둘러싸고 있는 흰색 콘크리트 진입로로 연장되어 건물의 입지를 강화시켜준다.
1층과 지하의 중심부를 관통하는 뚫린 공간은 다리로 연결되어 있다. 1층 주 출입구를 통해 글래스 큐브로 들어가는 방문객들은 수평뿐만 아니라 위 아래로도 트여 있는 공간과 만나게 된다. 두 층에 걸쳐 말려들어간 벽은 움푹 들어간 공간을 형성하여 주제별 제품 전시장과 회의실 등 다양한 용도로 이용된다. 특히, 벽을 분절하는 선들은 중요한 그래픽 디자인 요소인 동시에 천장으로 이어져 환기 시설의 접합부가 된다. 입면을 마주보고 있는 흰 벽은 앞에 거즈를 깔아 놓은 듯이 시각적으로 녹아든다. 햇빛과 역동적으로 구성된 인공 조명은 순백색 내부에 하이라이트를 주며 끊임없이 분위기를 변화시킨다.

Architect 3deluxe in/exterior **Graphic design** 3deluxe graphics **Project development and management** system modern GmbH **Construction management, realisation, statics** Ingenieurburo J. Steinkemper GmbH **Facade planning** Schlaich Bergermann und Partner GmbH **Facade construction** Metallbau Renneke **Dry-construction** Laackmann Trockenbau **'Genetics' and pilaster strips on the facade** Rosskopf & Partner AG **Electrical planning** Veverka & Druke Ingenieurgesellschaft mbH **Building services, ventilation planning** Pa-Bra GmbH & Co **Light planning** Licht und Gestaltung Schiefer GmbH **Outdoor path network** Shapers KG **Furnishing, Finishing** Kessler Innenausbau, Rosskopf & Partner AG, Hillebrand GmbH **Textile interior furnishing** Handelskontor Raumausstattung **Paintworks and carpeting**: Walecki Maler und Lackierer **Client**: Leonardo / Glaskoch B. Koch Jr. GmbH **Location**: Bad Driburg **Expanse** 2.900 m^2 **Enclosed space** 18.500 m^2 **Start of planning** 2004 **Completion** 2007 **Photographs** Courtesy of the architect ©Emanuel Raab, 3deluxe - p.170

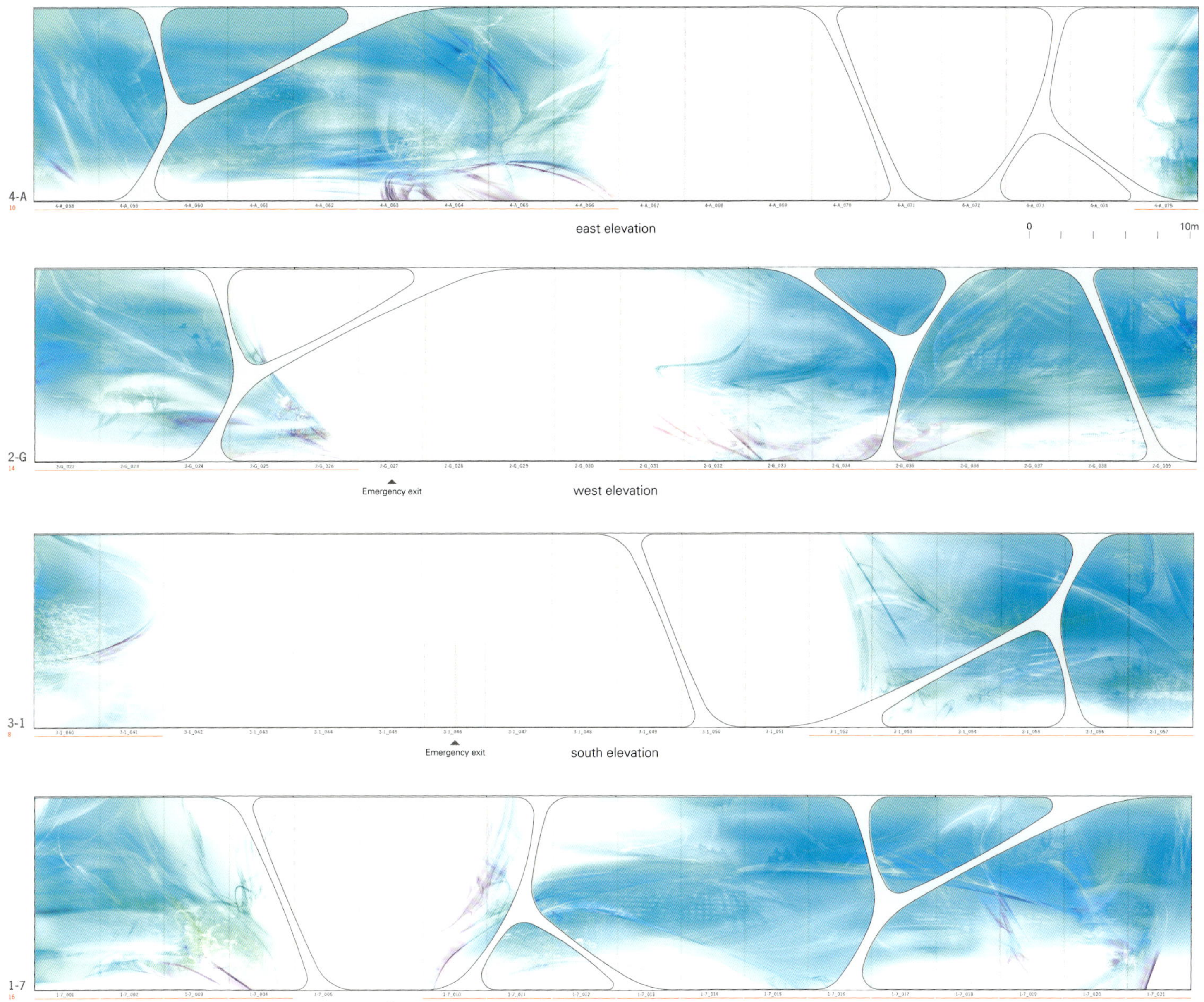
4-A
east elevation
0
10m
2-G
Emergency exit
west elevation
3-1
Emergency exit
south elevation
1-7
Main entrance
north elevation

elevation: glass panels no. 64 and 65 (east facade)

elevation: glass panels no. 14 and 15 (north facade)

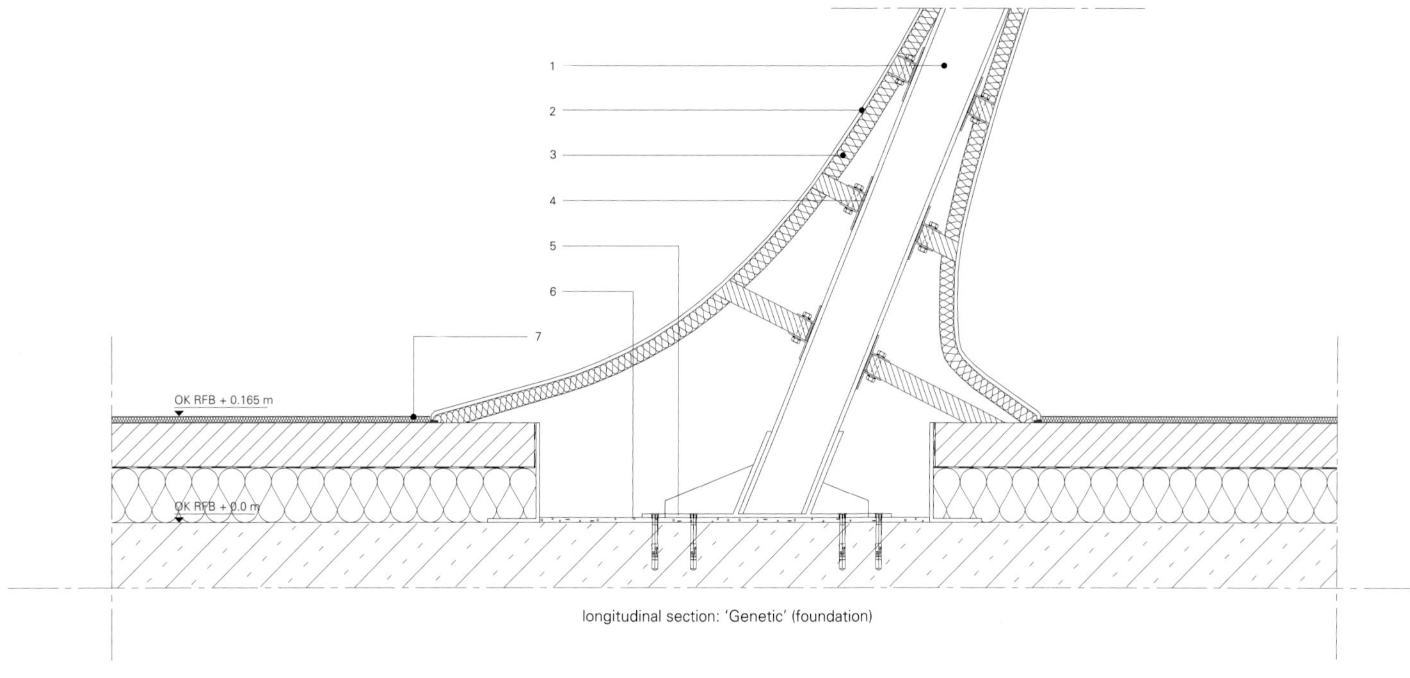

longitudinal section: 'Genetic' (foundation)

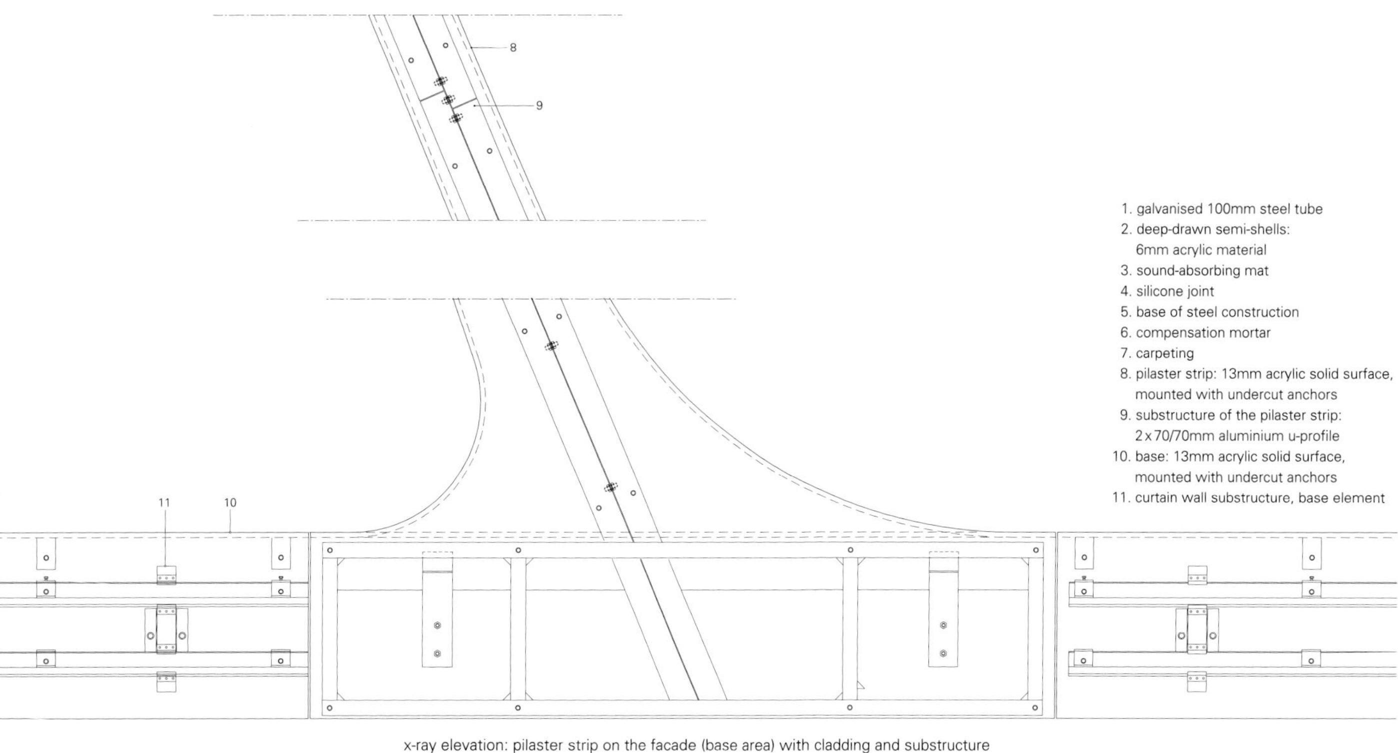

x-ray elevation: pilaster strip on the facade (base area) with cladding and substructure

1. galvanised 100mm steel tube
2. deep-drawn semi-shells: 6mm acrylic material
3. sound-absorbing mat
4. silicone joint
5. base of steel construction
6. compensation mortar
7. carpeting
8. pilaster strip: 13mm acrylic solid surface, mounted with undercut anchors
9. substructure of the pilaster strip: 2 x 70/70mm aluminium u-profile
10. base: 13mm acrylic solid surface, mounted with undercut anchors
11. curtain wall substructure, base element

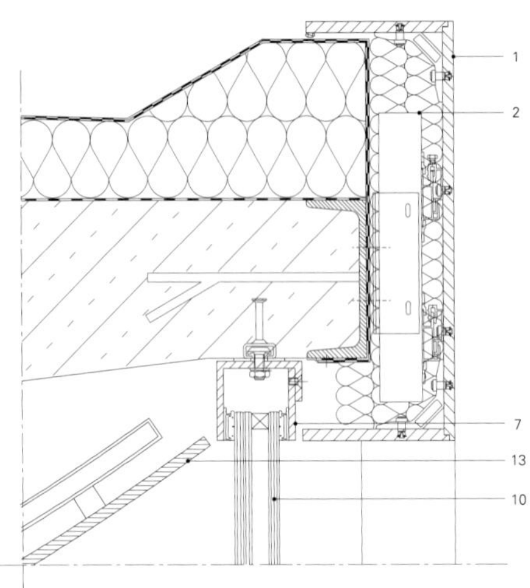

longitudinal section: attic

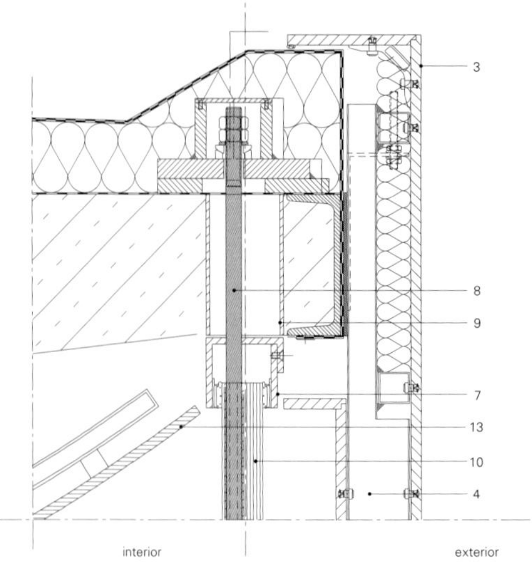

longitudinal section:
pilaster strips on the facade and upper tension anchor

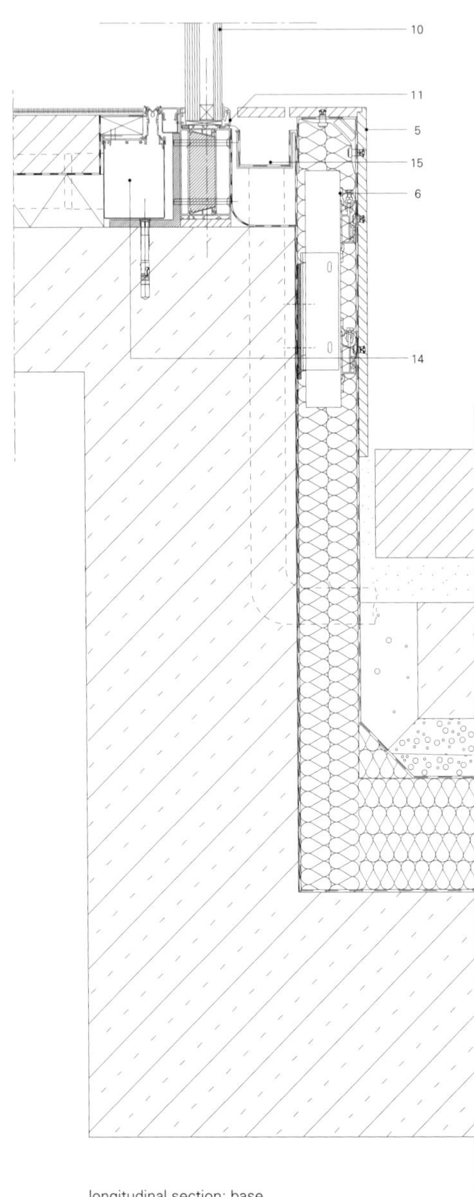

longitudinal section: base

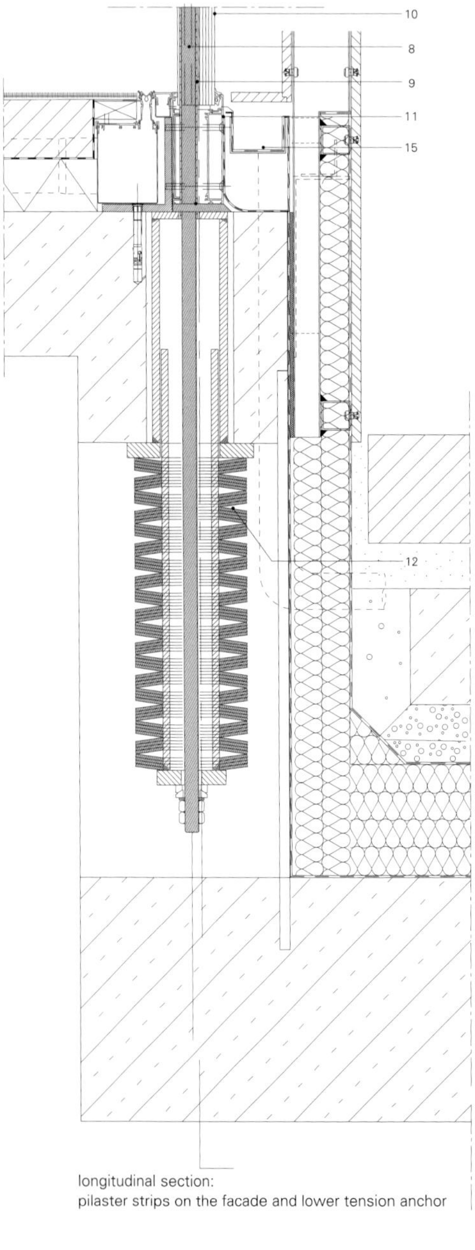

longitudinal section:
pilaster strips on the facade and lower tension anchor

1. cladding of the attic: 13mm acrylic solid surface, mounted with undercut anchors
2. curtain wall substructure, attic element
3. pilaster strip: 13mm acrylic solid surface, mounted with undercut anchors
4. substructure of the pilaster strip: 2 x 70/70mm aluminium u-profile
5. base: 13mm acrylic solid surface, mounted with undercut anchors
6. curtain wall substructure, base element
7. upper frame profile: aluminium
8. 20mm tension rod, stainless steel
9. 25/25/2mm square profile, stainless steel
10. 5875/1970mm insulation glass 70/40 interior: LSG of two 10mm float glass panels, one laminated with printed PVB film exterior: 12mm float glass
11. lower frame profile: aluminium
12. disk spring column with 19 sets of 4 single springs
13. suspended ceiling, bent plasterboard
14. floor ventilation outlet
15. drainage gutter

cross section: glass facade

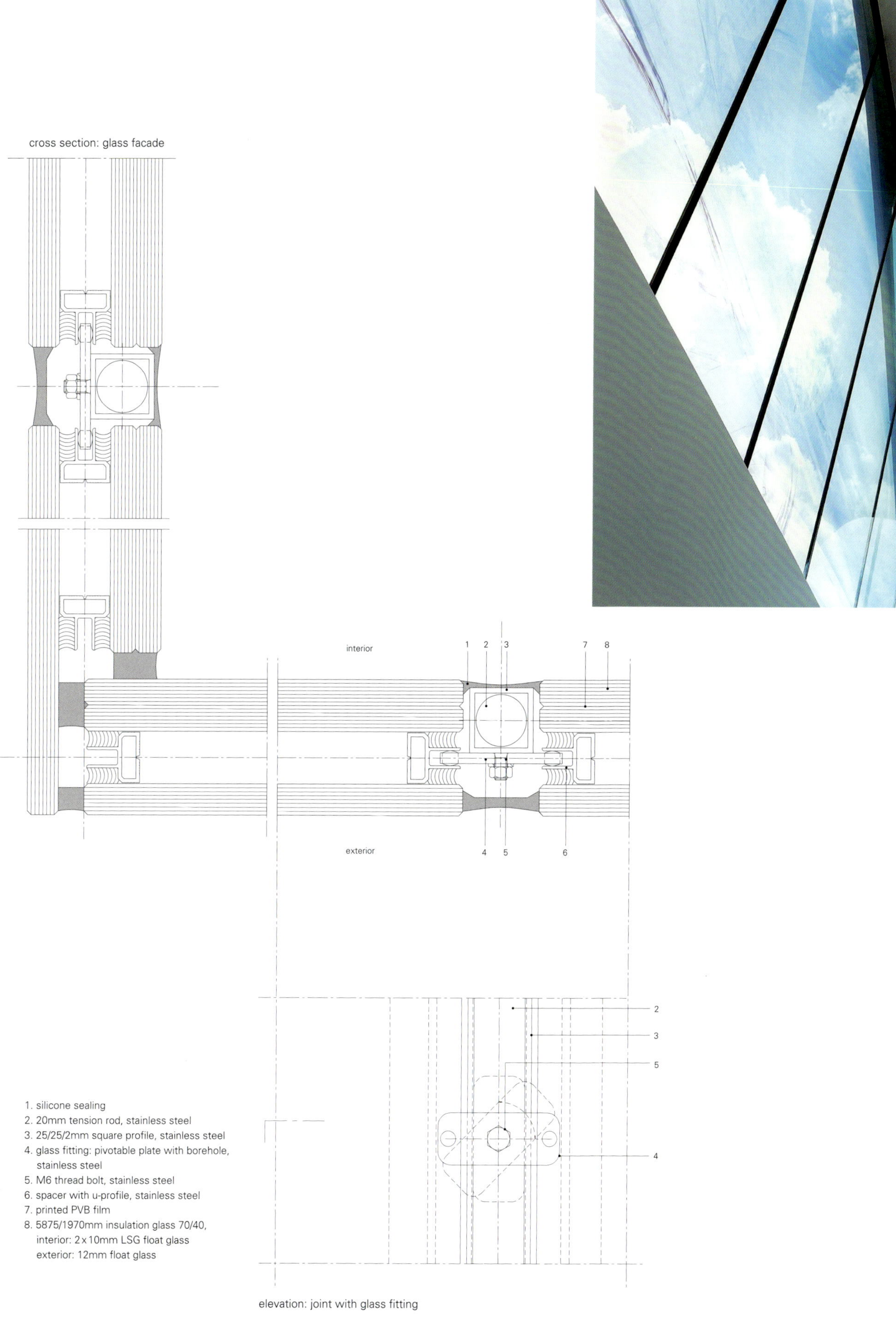

1. silicone sealing
2. 20mm tension rod, stainless steel
3. 25/25/2mm square profile, stainless steel
4. glass fitting: pivotable plate with borehole, stainless steel
5. M6 thread bolt, stainless steel
6. spacer with u-profile, stainless steel
7. printed PVB film
8. 5875/1970mm insulation glass 70/40, interior: 2x10mm LSG float glass exterior: 12mm float glass

elevation: joint with glass fitting

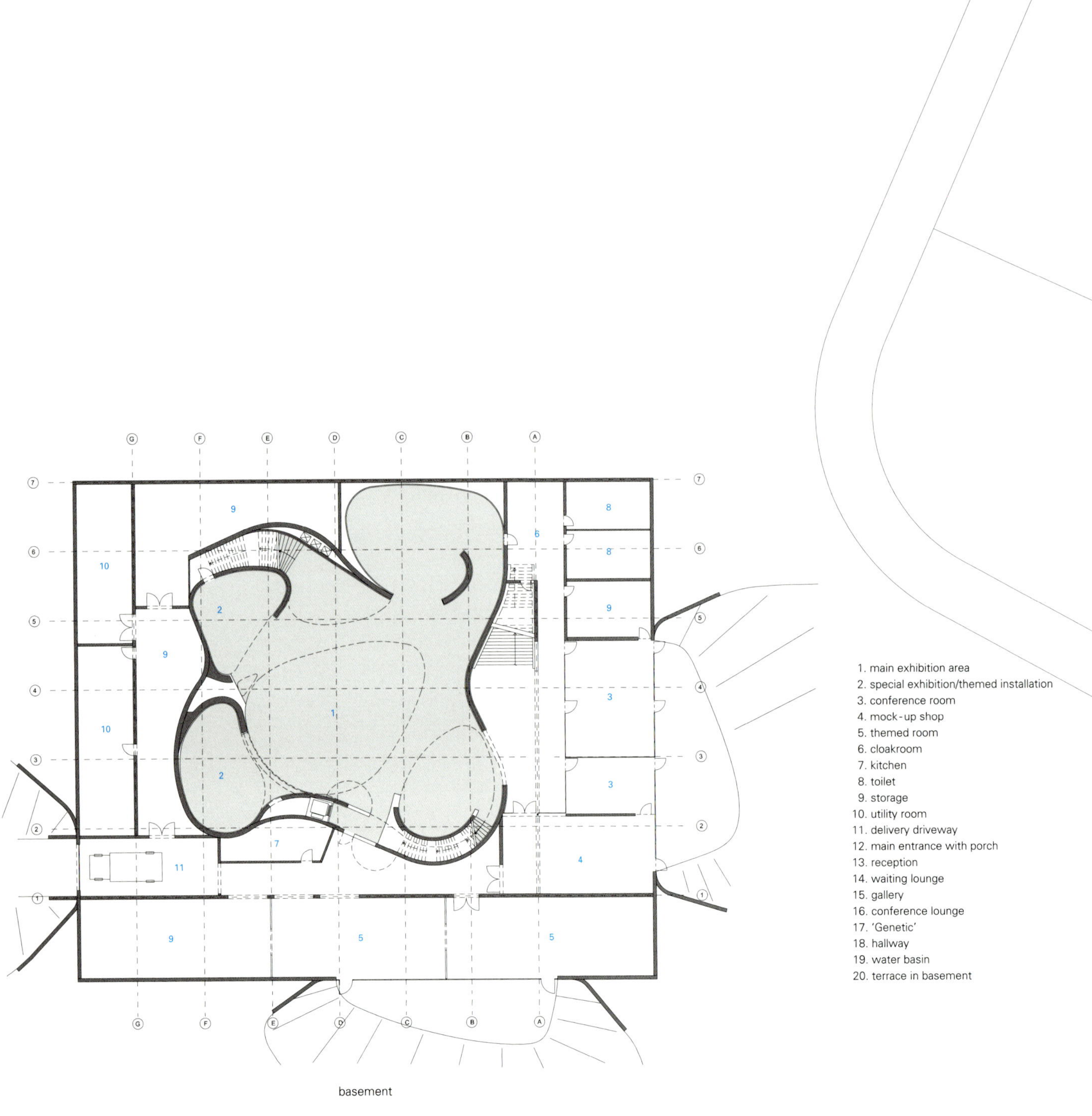

1. main exhibition area
2. special exhibition/themed installation
3. conference room
4. mock-up shop
5. themed room
6. cloakroom
7. kitchen
8. toilet
9. storage
10. utility room
11. delivery driveway
12. main entrance with porch
13. reception
14. waiting lounge
15. gallery
16. conference lounge
17. 'Genetic'
18. hallway
19. water basin
20. terrace in basement

basement

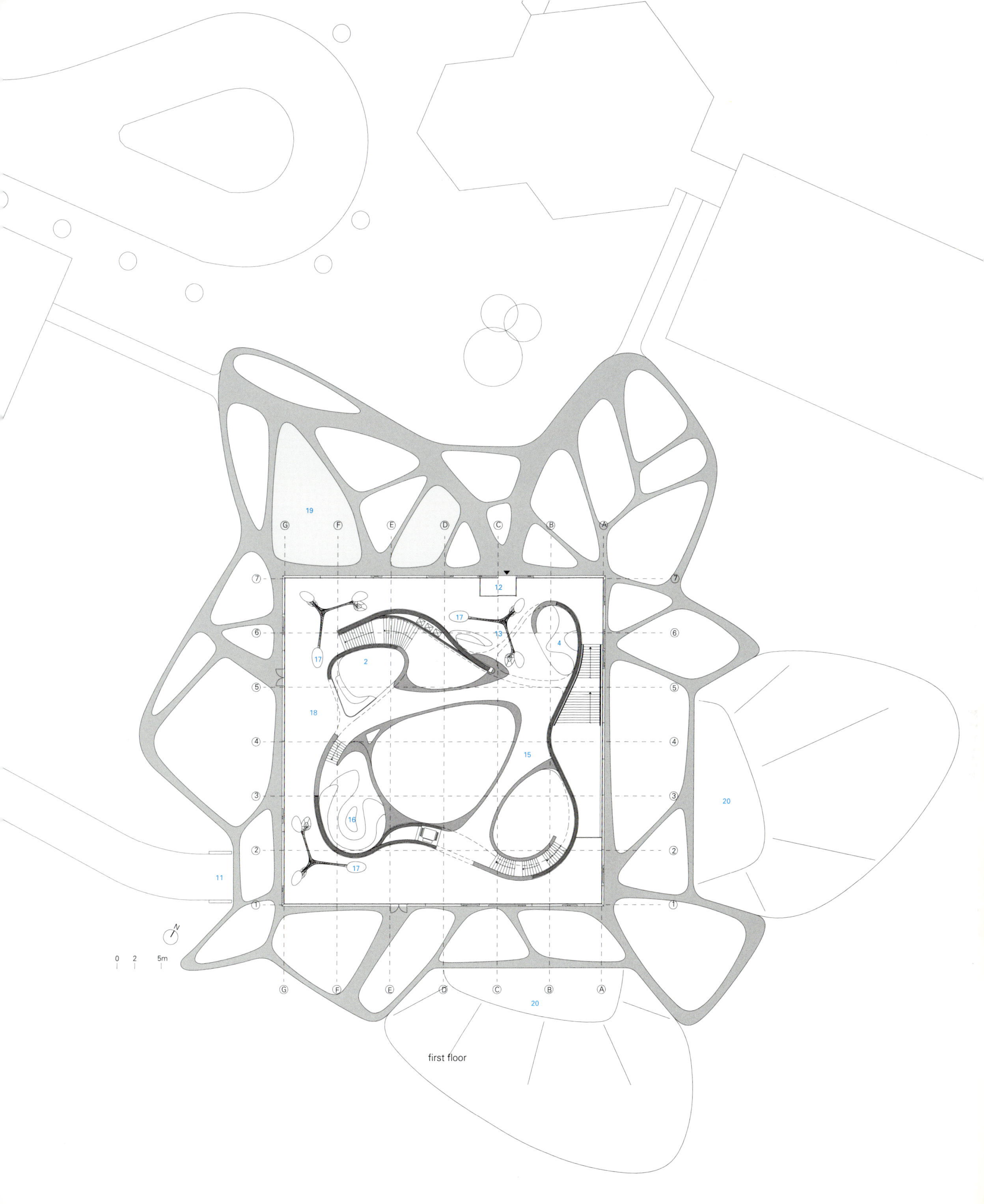

first floor

1. 'Genetic'
2. pilaster strip on the facade
3. porch/main entrance
4. dry-construction wall with gauze curtain
5. hallway
6. utility room
7. storage
8. gallery
9. main exhibition area

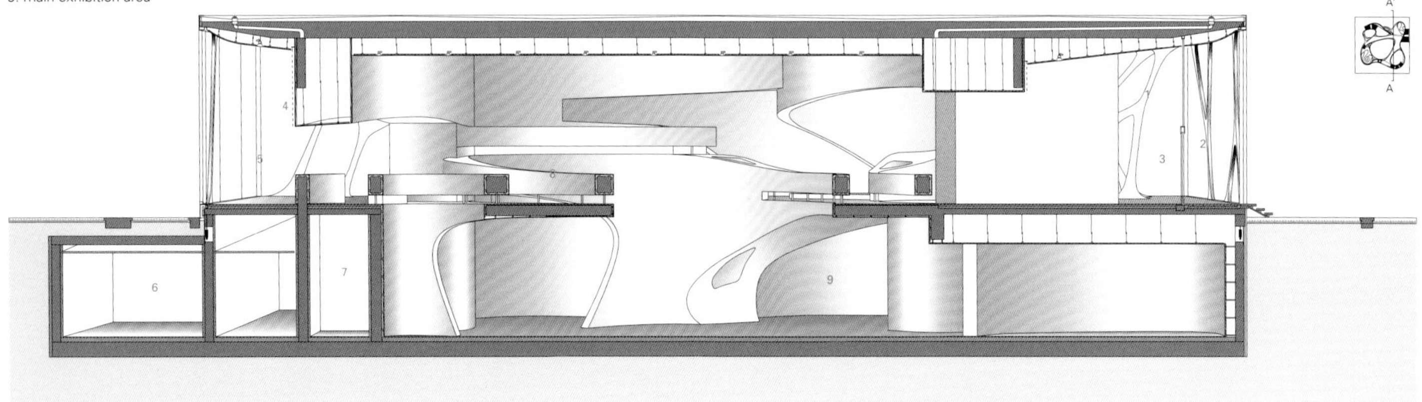

section A - A'

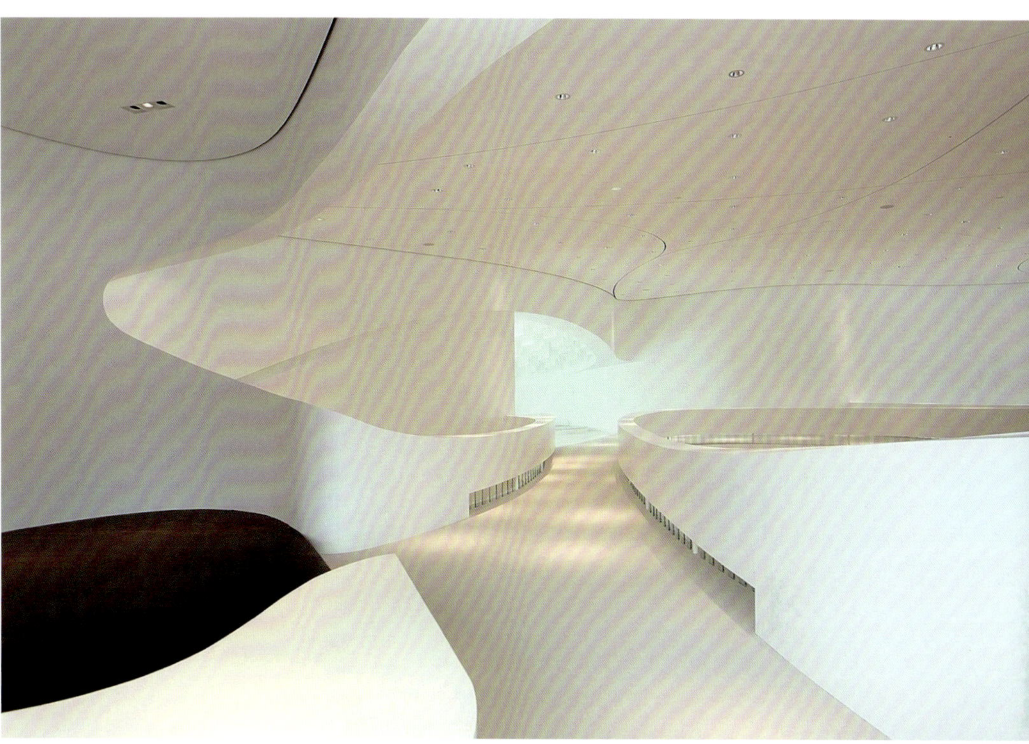

1. suspended ceiling: plasterboard panels
2. gauze curtain
3. grid of measurement points: 0.3 m

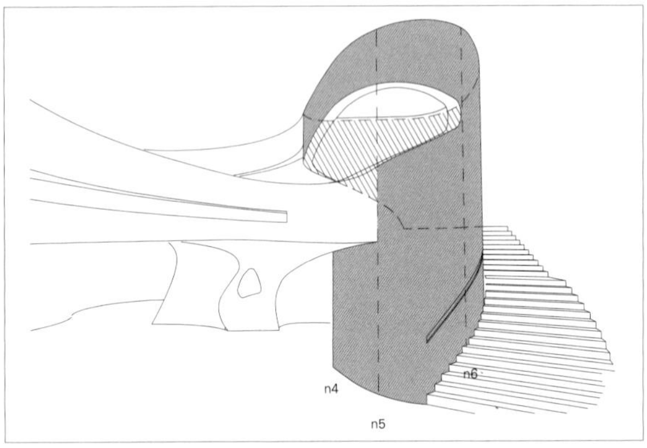

perspective view: basement

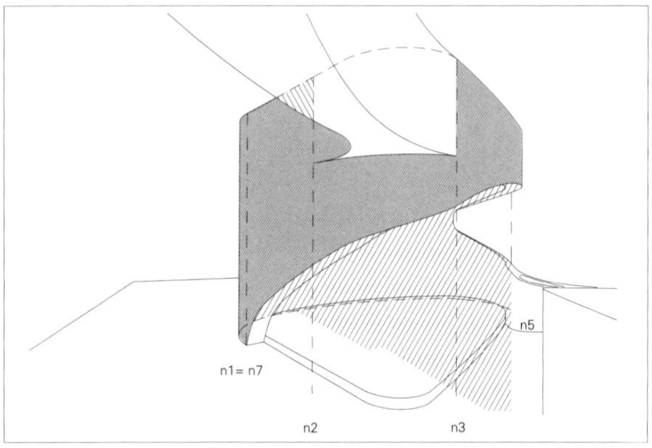

perspective view: first floor

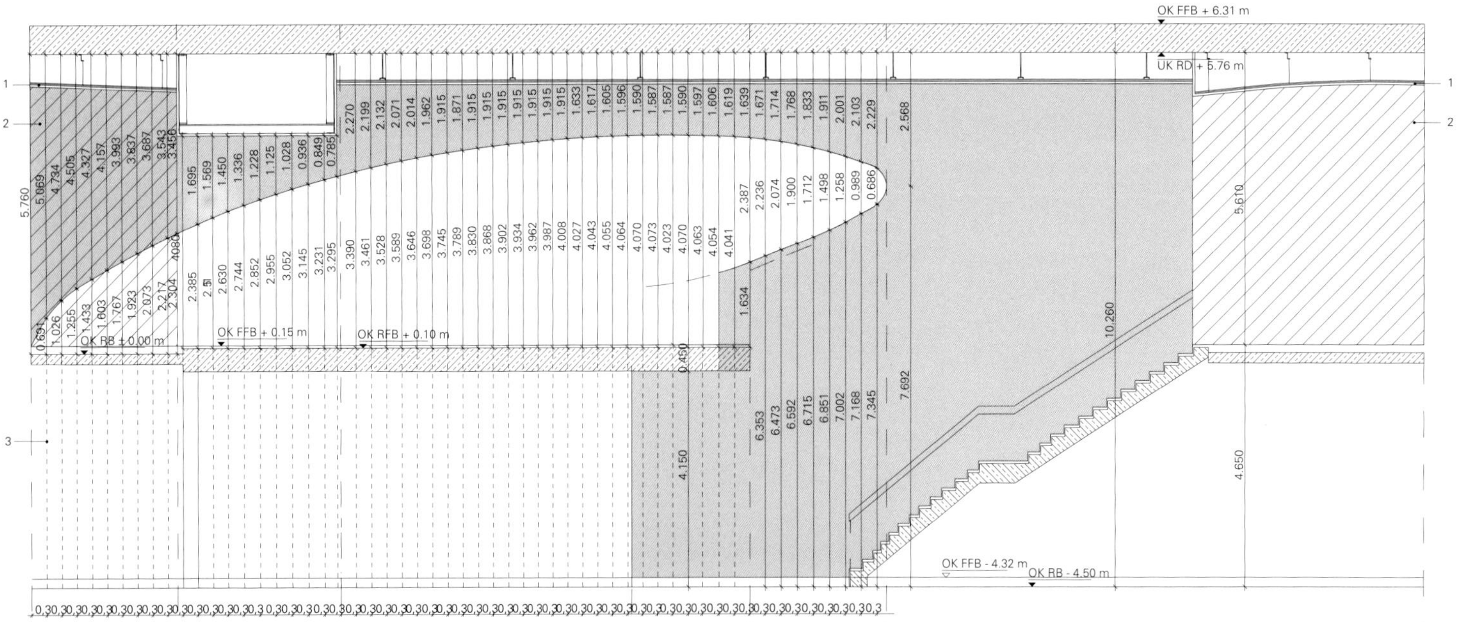

full-length projection: dry-construction wall

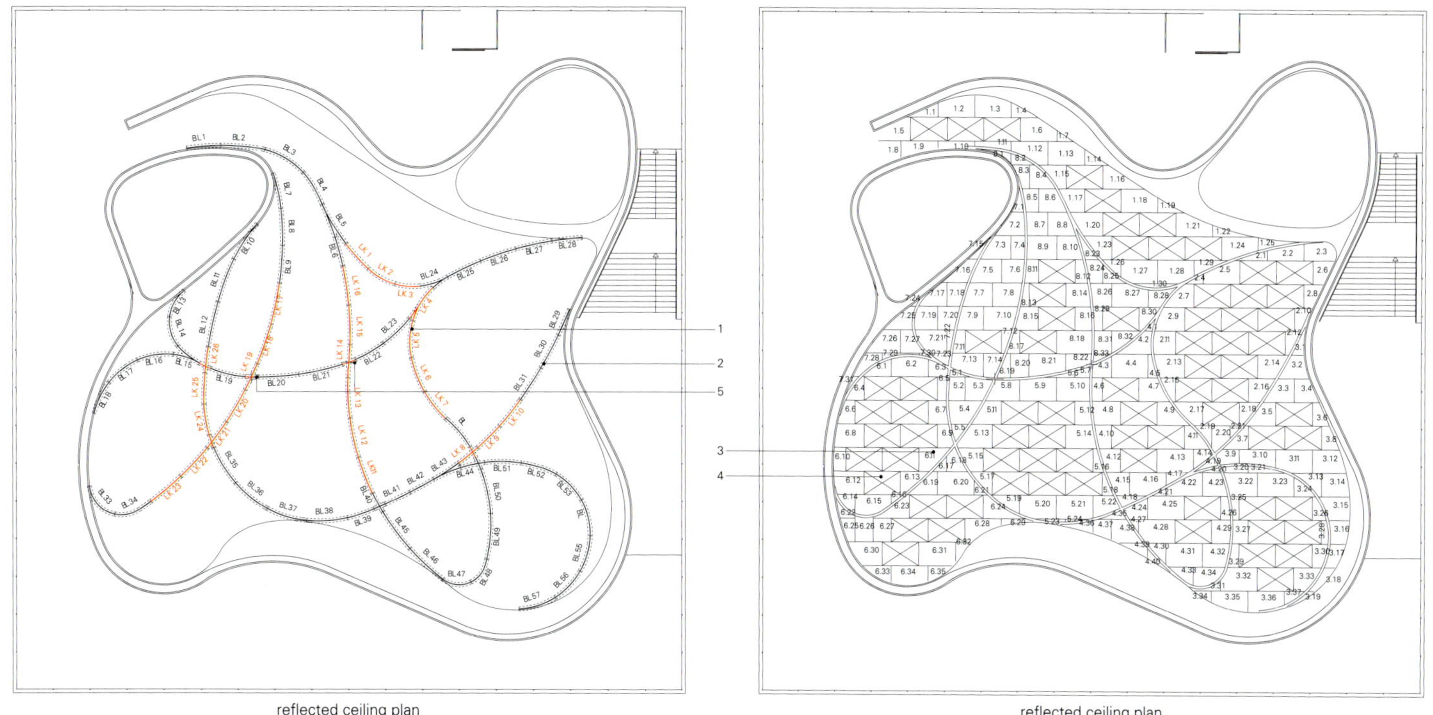

reflected ceiling plan
ventilation pipes and blind joints

reflected ceiling plan
laying plan: ceiling panels

1. ventilation pipe prefabricated from plasterboard elements
2. blind joint prefabricated from plasterboard elements
3. prefabricated plasterboard connection panels
4. rectangular standard plasterboard panels
5. measurement points

1. inlet pipe
2. flexible connection tube
3. ventilation pipe prefabricated from plasterboard elements
4. prefabricated plasterboard connection panel
5. prefabricated plasterboard blind joint

longitudinal section: suspended ceiling with ventilation pipes and blind joints

Initially we proposed to place entire volume of the library underground to achieve a space resembles a cave. However, for various reasons we found that the proposal was impossible. Although we start designing the volume above the ground, we still tried to build a subterranean space on the ground. We designed a continuous dome-like space by subtracting volumes from the "subterranean space above the ground". We trimmed down more volumes and it became a structural system of a series of domes and arches.

The arches have been designed to follow gentle curves at different angles. These continuous curves of arches articulate space into blocks of squares and triangles. Due to the strategic placement of the furniture that penetrates the space, paradoxically contradicting characteristics were attributed to the reading area; flow and stand. The slope of the ground floor follows the natural decline of the land so that the architecture is well integrated into the surrounding environment. In this way, spatial continuity is maintained between inside and outside space. This spatial concept will give freedom and richness to the architecture, receiving the natural energy that rises up from the earth and transforming it into an architectural order.

타마 예술 대학교의 도서관

처음에는 도서관 건물 전체를 지하에 두어 동굴 같은 공간을 만들고자 하였다. 그러나 여러 가지 사정으로 인해 그 계획은 불가능하다는 결론을 내렸다. 그래서 다시 지상에 건물을 설계하였지만, 여전히 지하와 같은 공간을 만들어내려고 하였다. 매스를 떼어내는 방식으로 연속하는 돔을 만들어 '지상에 있는 지하공간'을 설계하였으며, 매스들을 더 잘라내어 돔과 아치로 이루어진 구조를 완성하였다.

아치들은 여러 각도에서 부드러운 곡선을 그린다. 이 연속적인 곡선들은 공간을 사각형 또는 삼각형으로 분절시킨다. 이와 반대로 가구는 의도적으로 분절된 공간을 관통하도록 배치하여 독서 공간을 조성하였다. 이는 흐름과 정지의 상반된 특성을 나타낸다. 1층은 대지의 경사를 따라 기울여, 건물이 주위 환경과 하나가 되며 내부와 외부의 공간이 연속되도록 하였다. 이러한 공간 개념은 대지로부터 받은 자연에너지를 건축적으로 승화시키며, 도서관에 자유와 풍요로움을 가져다 줄 것이다.

Toyo Ito

Tama Art University Library

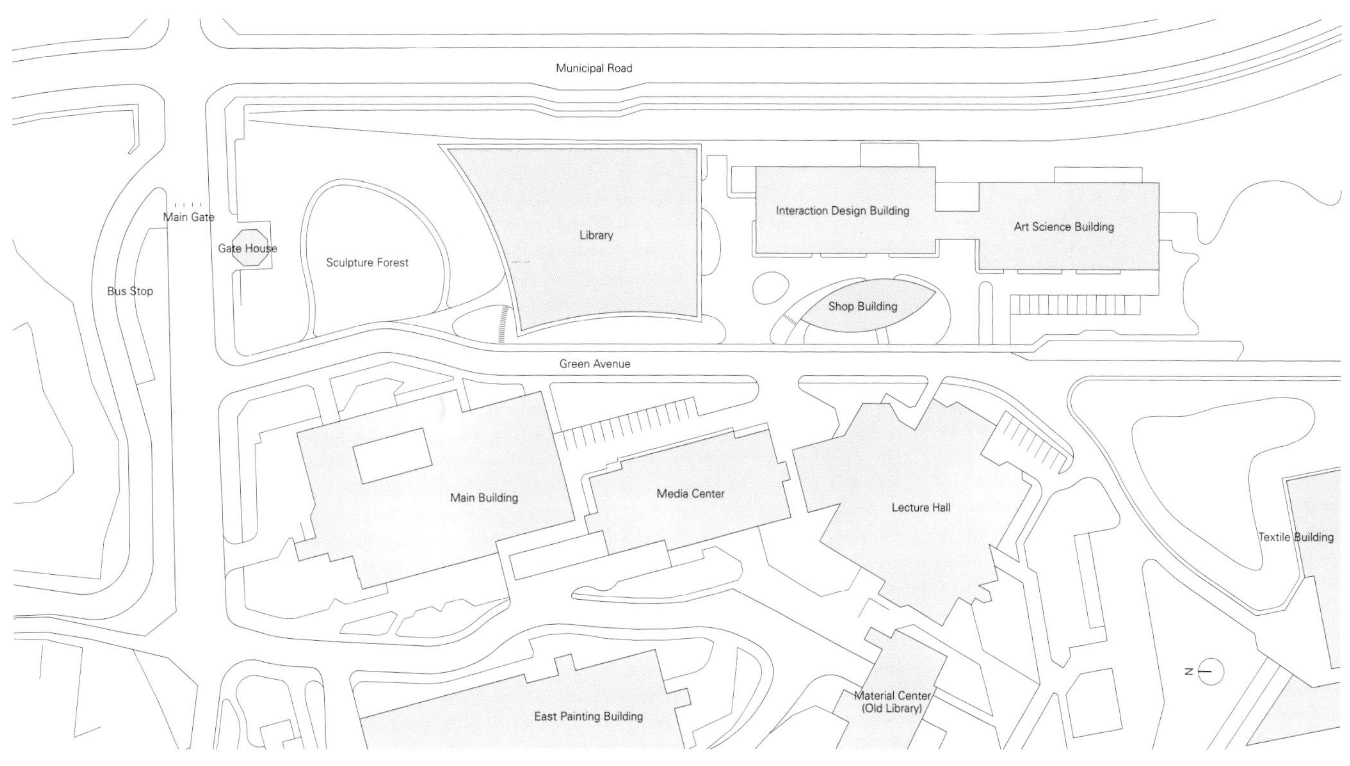

Architect Toyo Ito **Project team** Toyo Ito & Associates, Architects - Takeo Higashi, Hideyuki Nakayama, Yoshitaka Ihara **Associate architect** Kajima Design **Structural engineer** Sasaki Structural Consultants **Interaction design** Workshop for Architecture and Urbanism **Furniture design** Fujie Kazuko Atelier **Curtain design and fabricator** Nuno Corporation **General contractor** Kajima Corporation **Client** Tama Art University **Location** Hachioji City, Tokyo, Japan **Site area** 159,184.87 m^2 **Bldg. area** 2,224.59 m^2 **Gross floor area** 5,639.46 m^2 **Design period** 2004.4~2005.10 **Construction period** 2005.11~2007.2 **Photograph** Iwan Baan

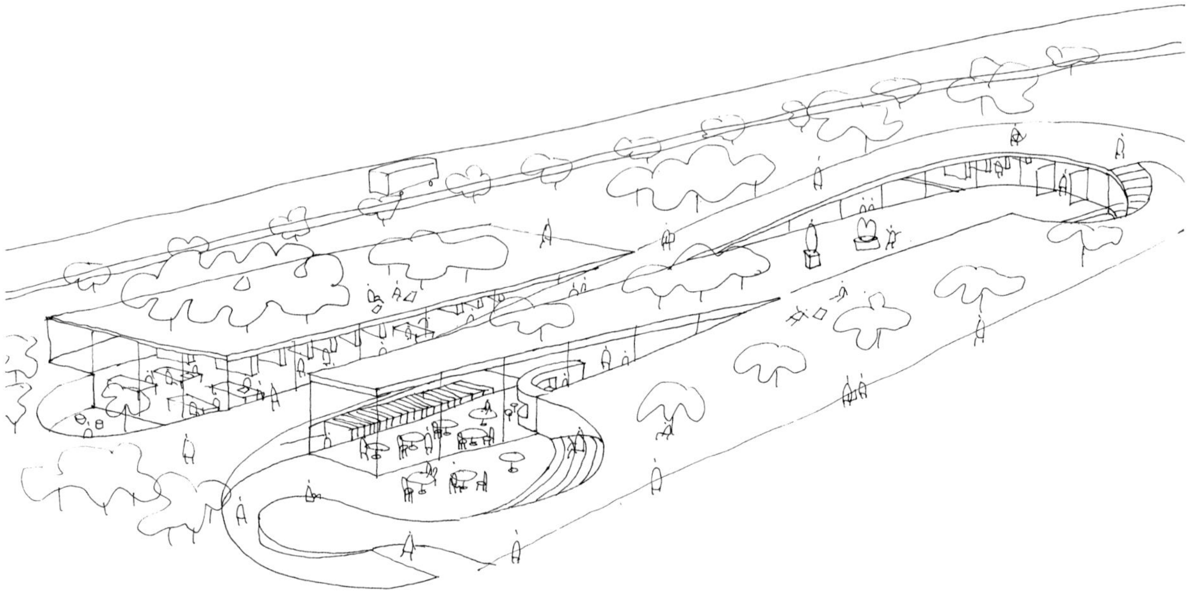

7000
6000
5000
3000
フラット
メディア ＋ 新刊
オフィス エリア
倉庫
搬入
1/20 スロープ
WC
検索
スロープ
新刊
2400
掲示
入口
BDS
受付
フラット
オフィスエリア
搬入口
集密書架
＋
倉庫
約 600 m²
土中
3.5M
5M
3M
CAFE
遮光カーテン
スロープ
TOPLIGHT
大型本
カーペット
2F オフィス
2F PLAN 041104

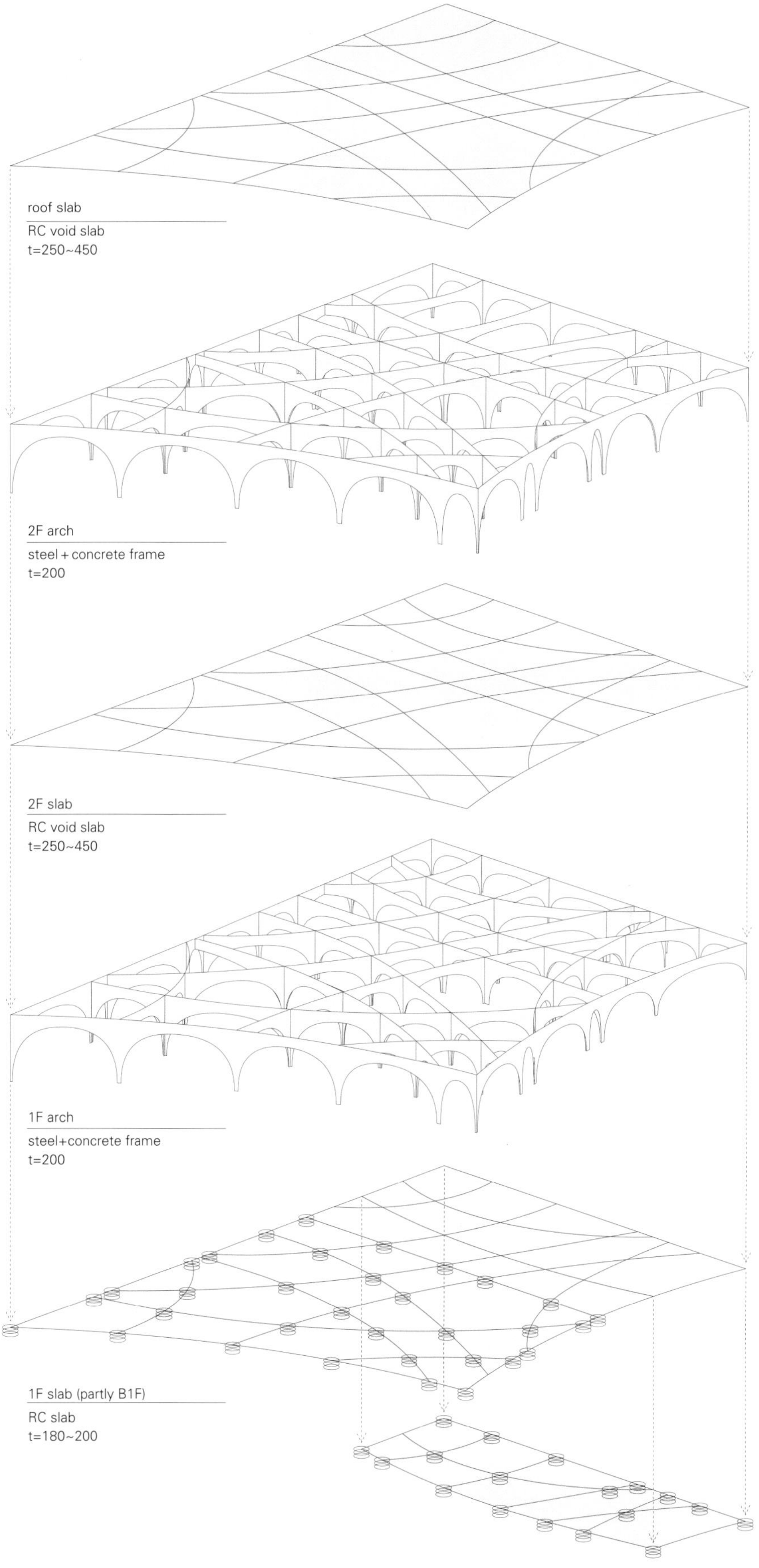

roof slab

RC void slab
t=250~450

2F arch

steel + concrete frame
t=200

2F slab

RC void slab
t=250~450

1F arch

steel+concrete frame
t=200

1F slab (partly B1F)

RC slab
t=180~200

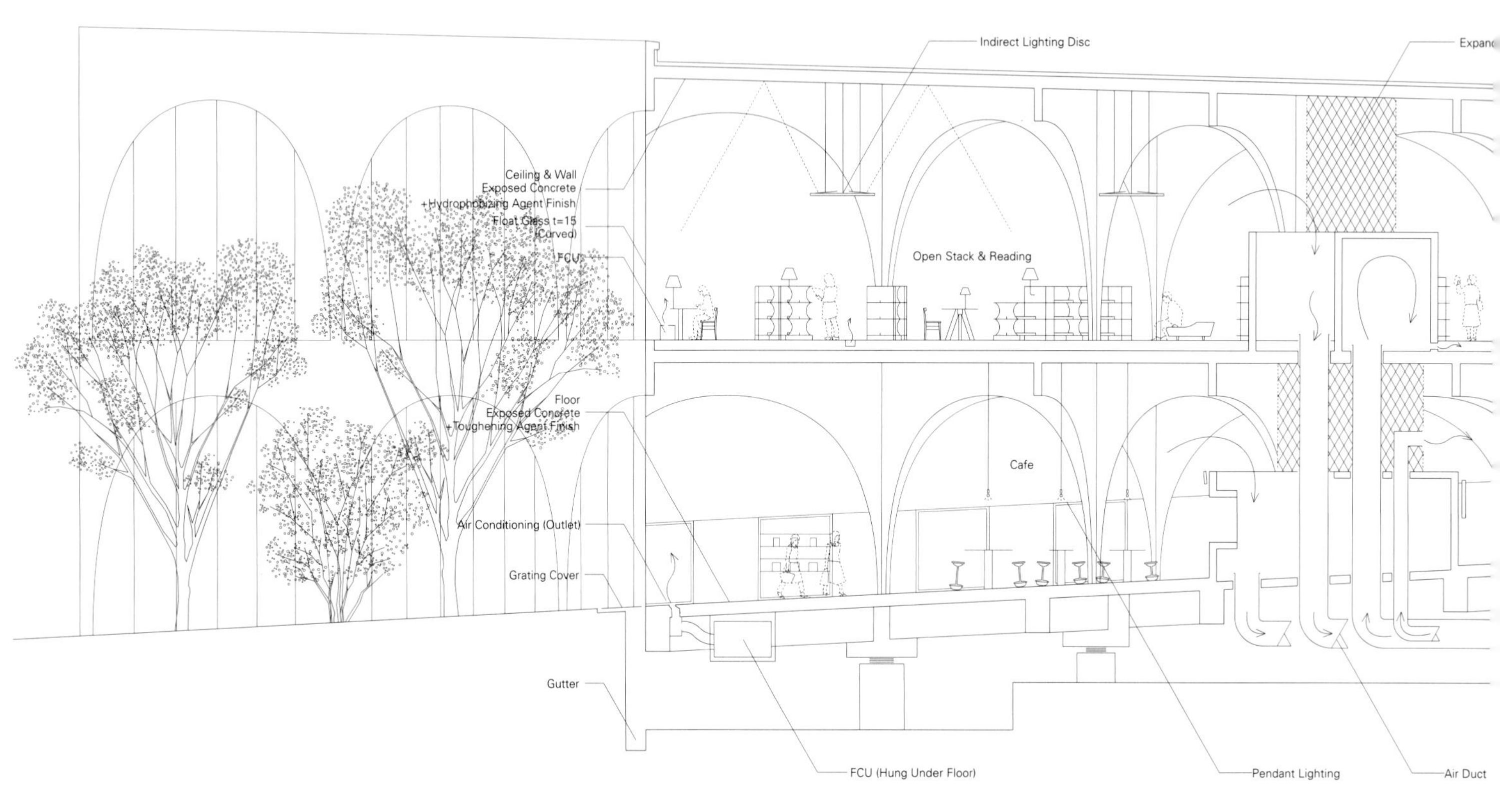

Indirect Lighting Disc
Expanc
Ceiling & Wall
Exposed Concrete
+Hydrophobizing Agent Finish
Float Glass t=15
(Curved)
FCU
Open Stack & Reading
Floor
Exposed Concrete
+Toughening Agent Finish
Cafe
Air Conditioning (Outlet)
Grating Cover
Gutter
FCU (Hung Under Floor)
Pendant Lighting
Air Duct

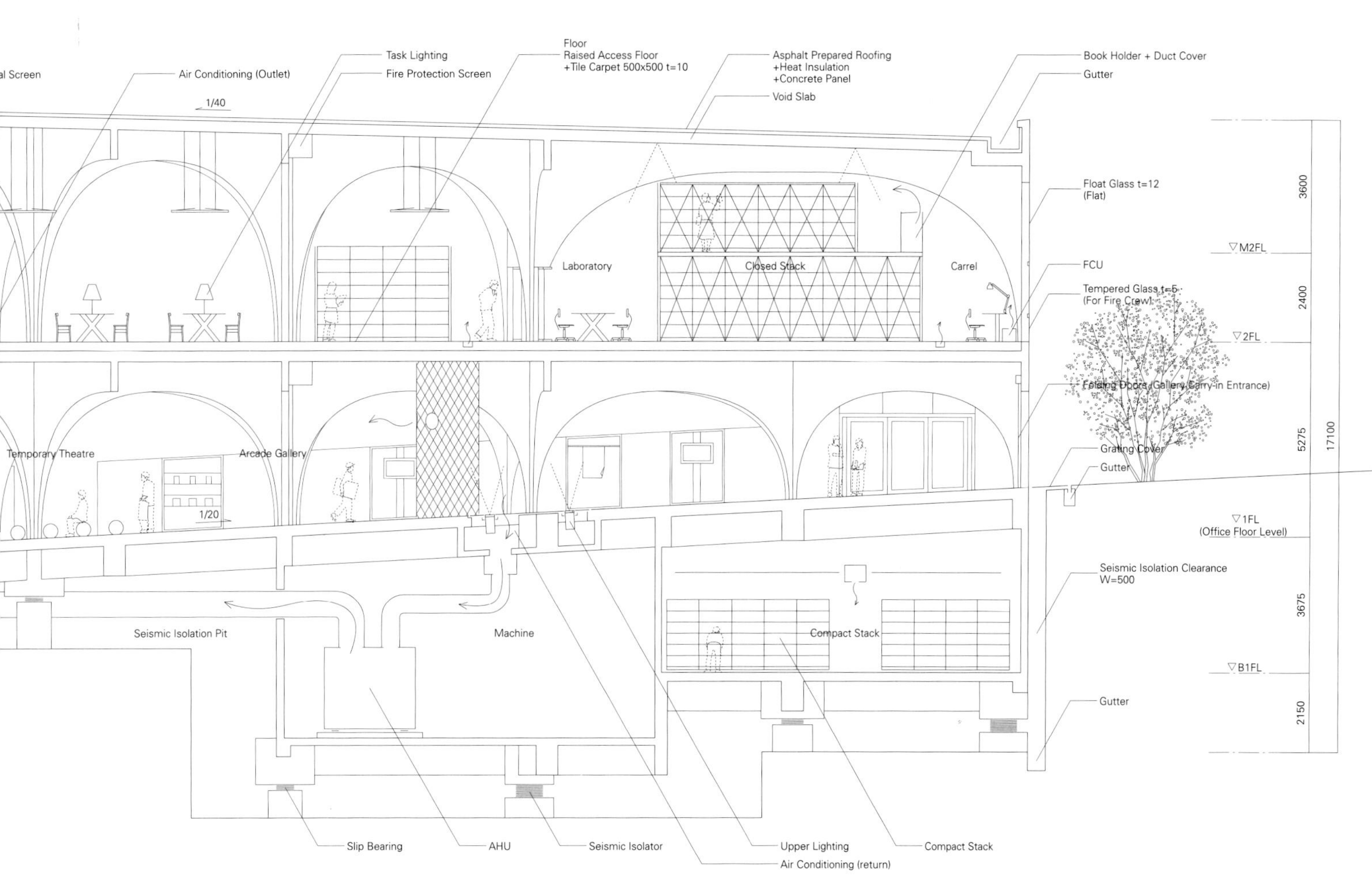

al Screen
Air Conditioning (Outlet)
Task Lighting
Fire Protection Screen
Floor
Raised Access Floor
+Tile Carpet 500x500 t=10
Asphalt Prepared Roofing
+Heat Insulation
+Concrete Panel
Void Slab
Book Holder + Duct Cover
Gutter
1/40
Float Glass t=12
(Flat)
3600
▽M2FL
Laboratory
Closed Stack
Carrel
FCU
Tempered Glass t=5
(For Fire Crew)
2400
▽2FL
Folding Doors (Gallery Carry-in Entrance)
Temporary Theatre
Arcade Gallery
1/20
Grating Cover
Gutter
5275
17100
▽1FL
(Office Floor Level)
Seismic Isolation Clearance
W=500
Seismic Isolation Pit
Machine
Compact Stack
3675
▽B1FL
Gutter
2150
Slip Bearing
AHU
Seismic Isolator
Upper Lighting
Air Conditioning (return)
Compact Stack

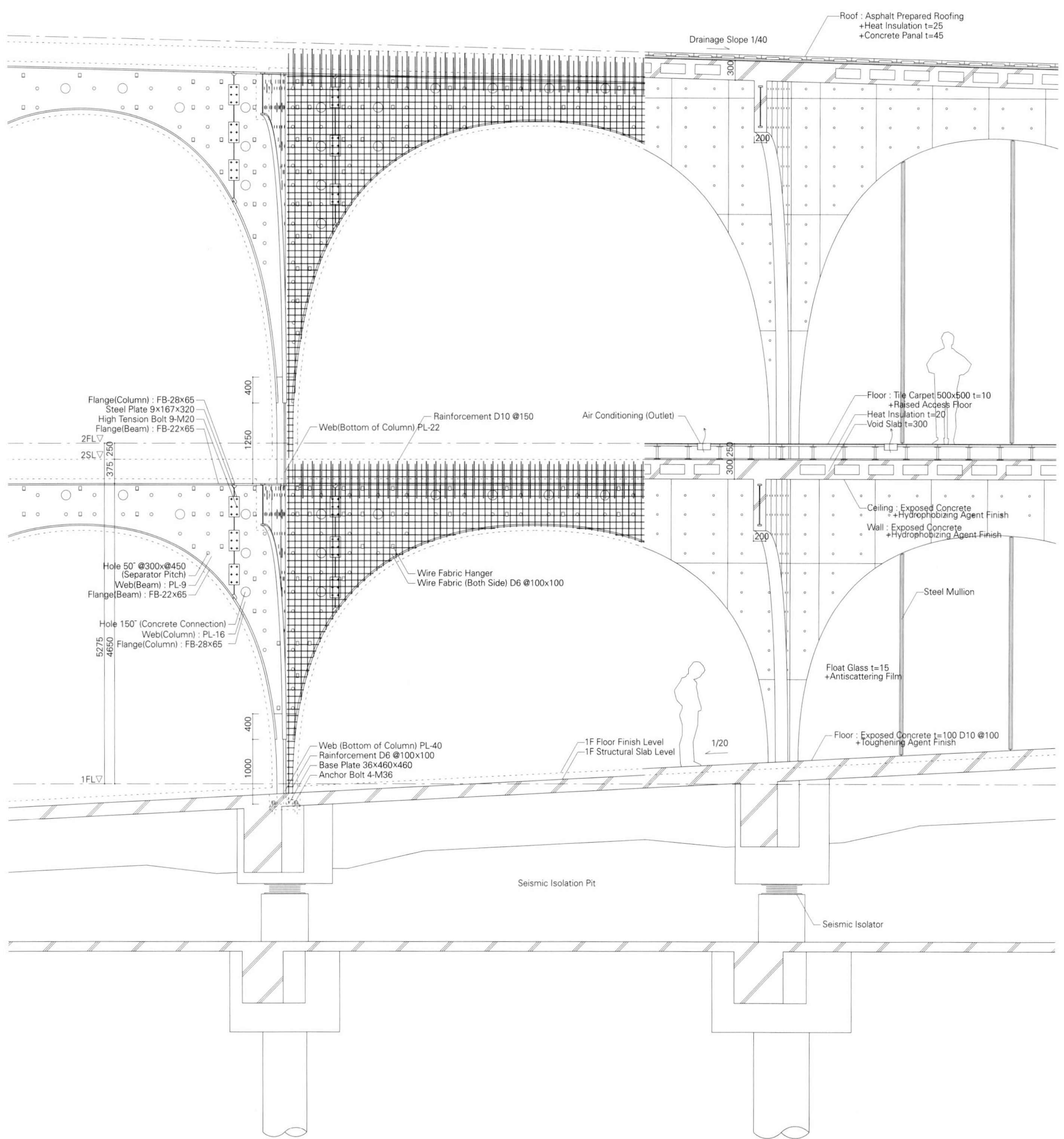

Roof : Asphalt Prepared Roofing
+Heat Insulation t=25
+Concrete Panal t=45
Drainage Slope 1/40
300
200
Flange(Column) : FB-28×65
Steel Plate 9×167×320
High Tension Bolt 9-M20
Flange(Beam) : FB-22×65
Rainforcement D10 @150
Web(Bottom of Column) PL-22
Air Conditioning (Outlet)
Floor : Tile Carpet 500x500 t=10
+Raised Access Floor
Heat Insulation t=20
Void Slab t=300
2FL
2SL
Ceiling : Exposed Concrete
+Hydrophobizing Agent Finish
Wall : Exposed Concrete
+Hydrophobizing Agent Finish
Hole 50˚ @300×@450
(Separator Pitch)
Web(Beam) : PL-9
Flange(Beam) : FB-22×65
Wire Fabric Hanger
Wire Fabric (Both Side) D6 @100×100
Steel Mullion
Hole 150˚ (Concrete Connection)
Web(Column) : PL-16
Flange(Column) : FB-28×65
Float Glass t=15
+Antiscattering Film
Web (Bottom of Column) PL-40
Rainforcement D6 @100×100
Base Plate 36×460×460
Anchor Bolt 4-M36
1F Floor Finish Level
1F Structural Slab Level
1/20
Floor : Exposed Concrete t=100 D10 @100
+Toughening Agent Finish
1FL
Seismic Isolation Pit
Seismic Isolator

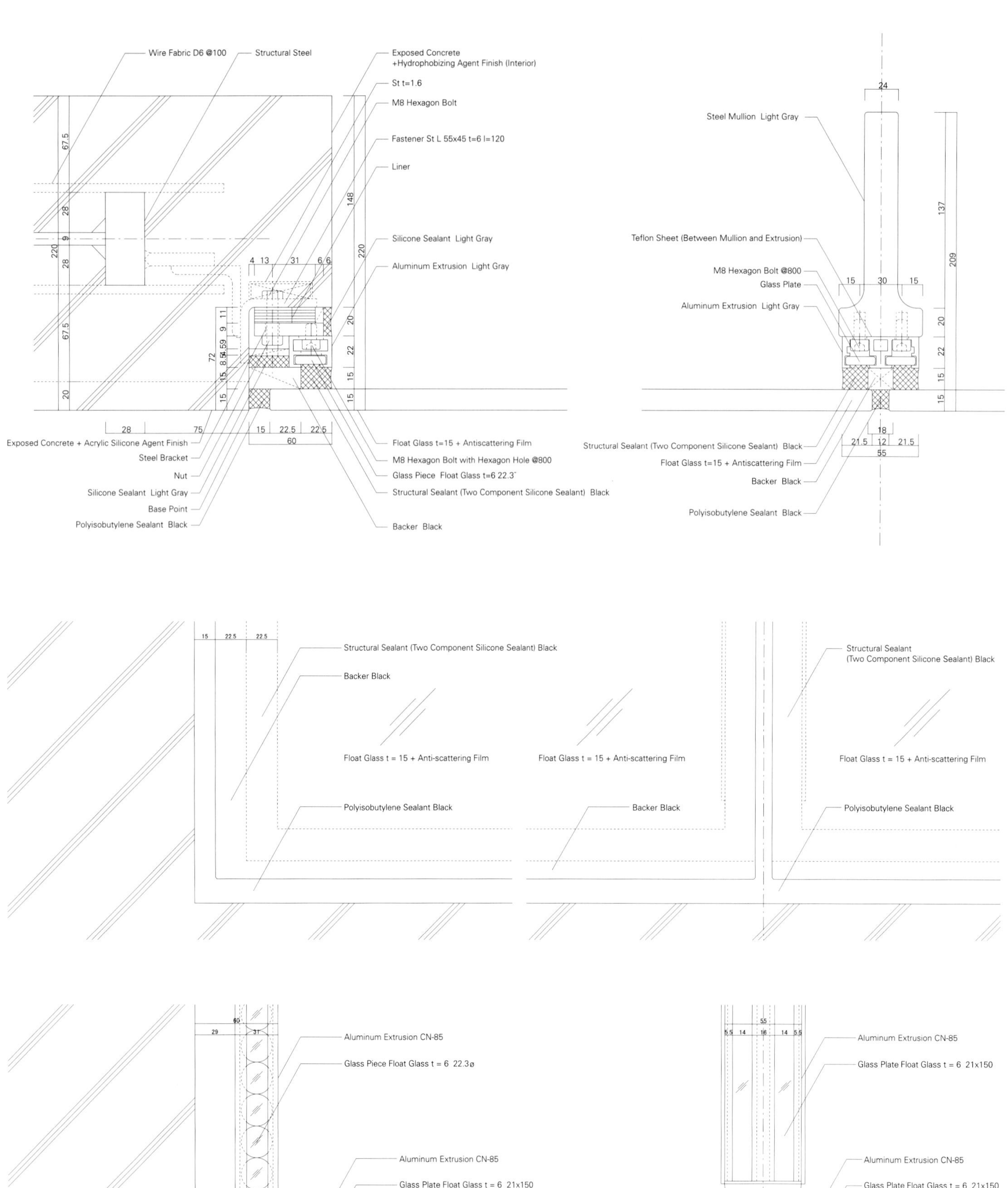
Wire Fabric D6 @100
Structural Steel
Exposed Concrete
+Hydrophobizing Agent Finish (Interior)
St t=1.6
M8 Hexagon Bolt
Fastener St L 55x45 t=6 l=120
Liner
Silicone Sealant Light Gray
Aluminum Extrusion Light Gray
Exposed Concrete + Acrylic Silicone Agent Finish
Steel Bracket
Nut
Silicone Sealant Light Gray
Base Point
Polyisobutylene Sealant Black
Float Glass t=15 + Antiscattering Film
M8 Hexagon Bolt with Hexagon Hole @800
Glass Piece Float Glass t=6 22.3
Structural Sealant (Two Component Silicone Sealant) Black
Backer Black
Steel Mullion Light Gray
Teflon Sheet (Between Mullion and Extrusion)
M8 Hexagon Bolt @800
Glass Plate
Aluminum Extrusion Light Gray
Structural Sealant (Two Component Silicone Sealant) Black
Float Glass t=15 + Antiscattering Film
Backer Black
Polyisobutylene Sealant Black
Structural Sealant (Two Component Silicone Sealant) Black
Backer Black
Float Glass t = 15 + Anti-scattering Film
Polyisobutylene Sealant Black
Float Glass t = 15 + Anti-scattering Film
Backer Black
Structural Sealant
(Two Component Silicone Sealant) Black
Float Glass t = 15 + Anti-scattering Film
Polyisobutylene Sealant Black
Aluminum Extrusion CN-85
Glass Piece Float Glass t = 6 22.3ø
Aluminum Extrusion CN-85
Glass Plate Float Glass t = 6 21x150
Aluminum Extrusion CN-85
Glass Plate Float Glass t = 6 21x150
Aluminum Extrusion CN-85
Glass Plate Float Glass t = 6 21x150

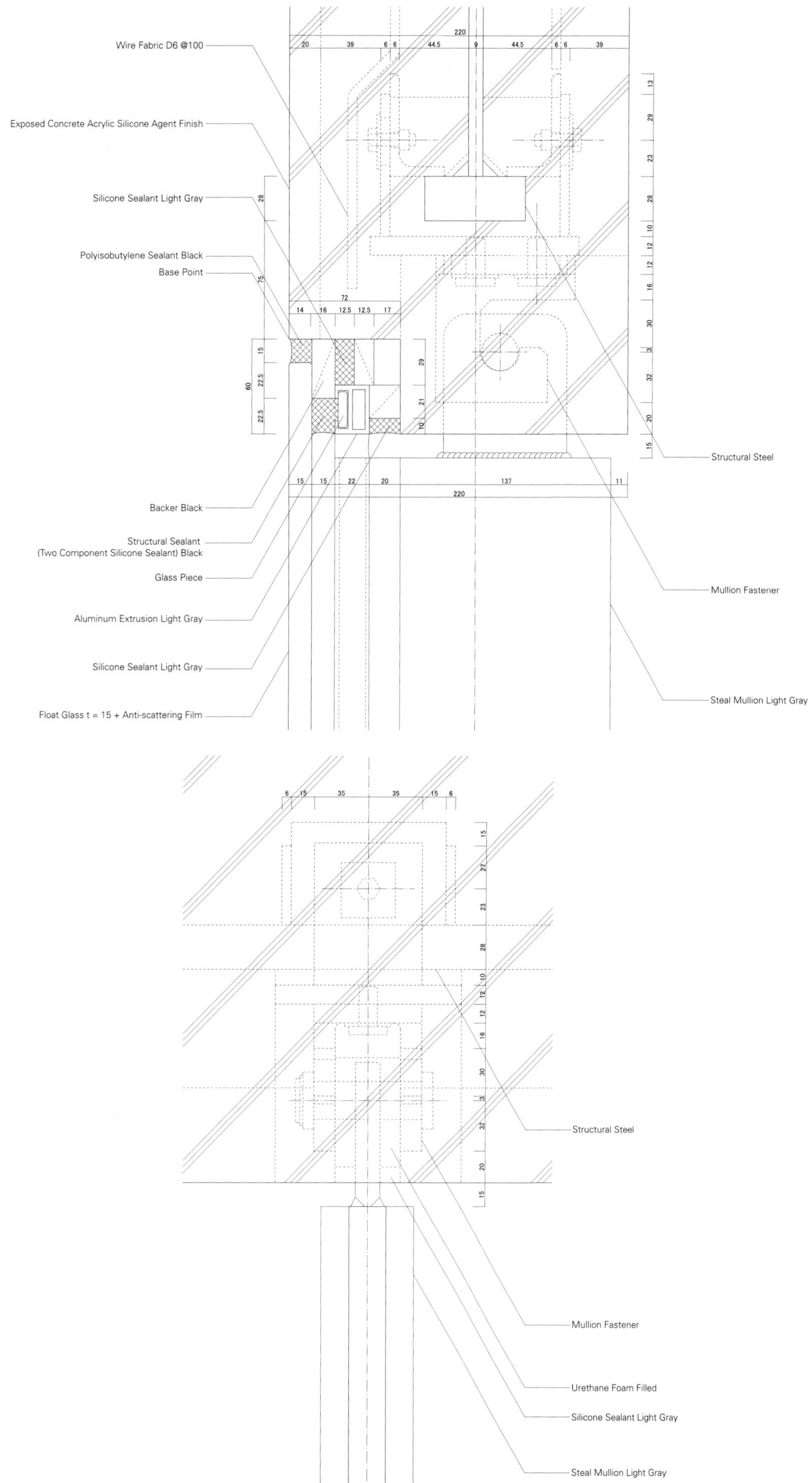
Wire Fabric D6 @100
Exposed Concrete Acrylic Silicone Agent Finish
Silicone Sealant Light Gray
Polyisobutylene Sealant Black
Base Point
Backer Black
Structural Sealant
(Two Component Silicone Sealant) Black
Glass Piece
Aluminum Extrusion Light Gray
Silicone Sealant Light Gray
Float Glass t = 15 + Anti-scattering Film
Structural Steel
Mullion Fastener
Steal Mullion Light Gray
Structural Steel
Mullion Fastener
Urethane Foam Filled
Silicone Sealant Light Gray
Steal Mullion Light Gray

コピー
CARD

12100
2400
11900
1300
1300
6000
4100
4100
6150
6150
6700
9100
4500
6400
0 2 4m

first floor

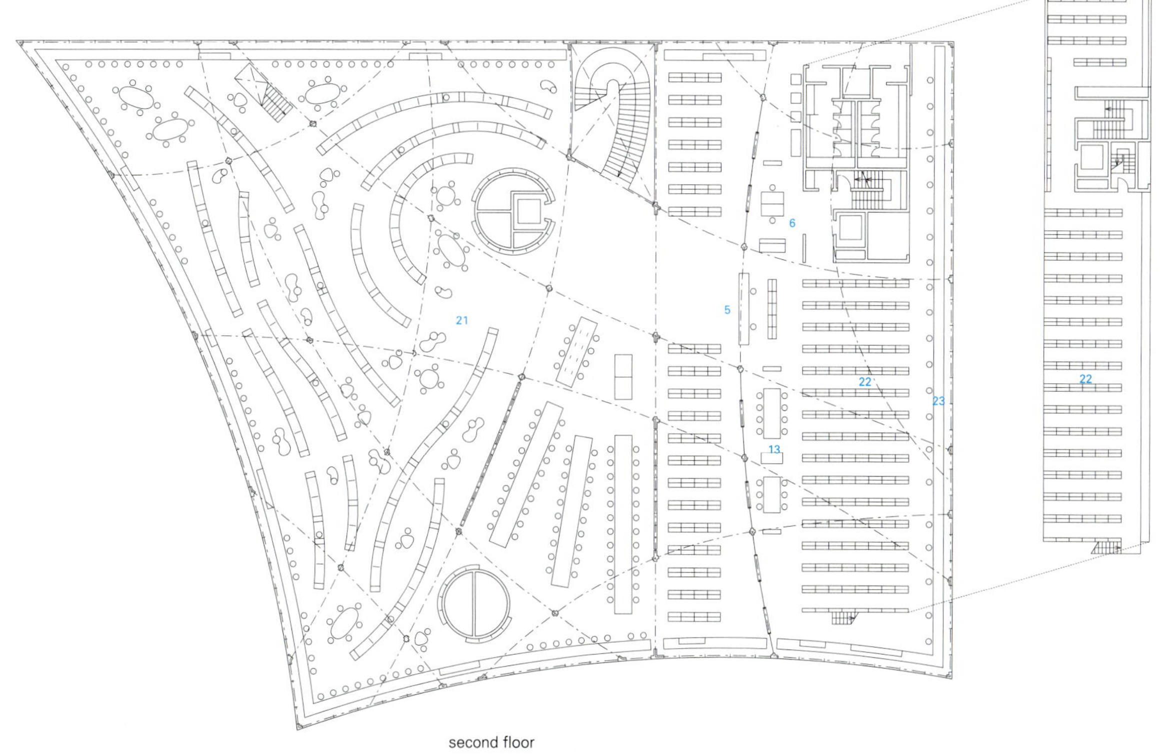

second floor

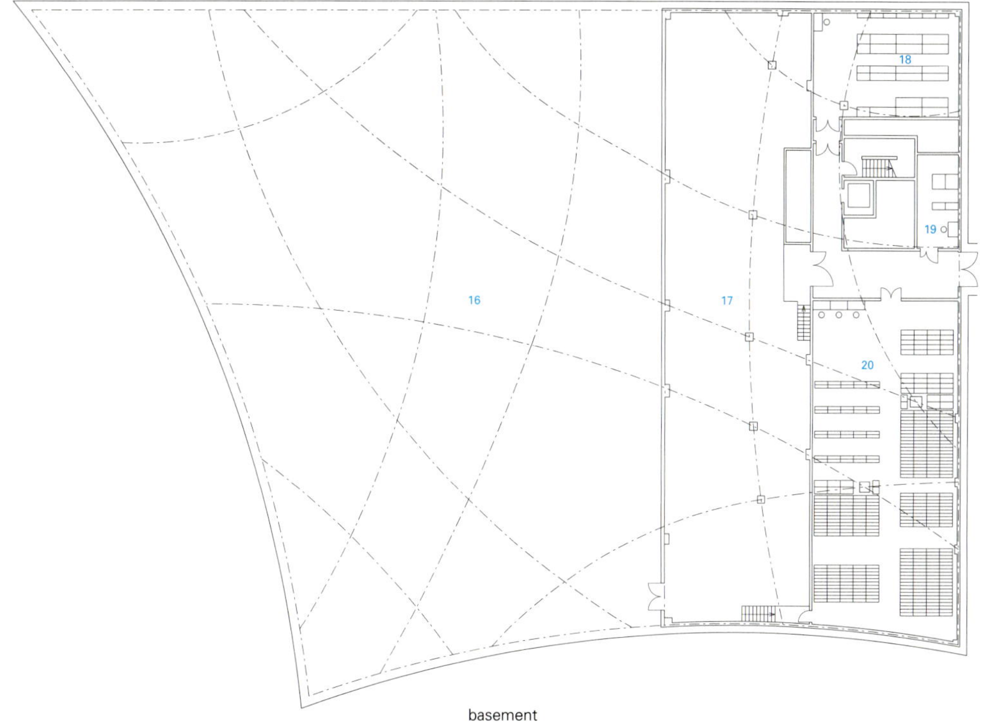

basement

1. lounge
2. AV booth
3. multimedia
4. new arrival magazine
5. information desk
6. office
7. library entrance
8. arcade gallery
9. temporary theatre
10. north entrance
11. cafe
12. locker
13. laboratory
14. south entrance
15. office entrance
16. seismic isolation pit
17. machine
18. valuable book stack
19. server
20. compact stack
21. open stack & reading
22. closed stack
23. carrel

RACE
FOR
THE
CURE

In this building we wanted to create well-ventilated offices with natural light to take advantage of the mild weather of Guadalajara. The design makes air-conditioning unnecessary. The client wished for a singular structure because the site is located in a dense office district.
The area was fixed at 4,800 square meters, but the height of the building was left to us. We decided on a high-rise building developed from three central concrete cores, which contain all the facilities and vertical circulation. These cores work as columns and are the only structural support for the building. Great-cantilevered beams of various dimensions support post-stressed, column-free slabs. Thus both office modules and the parking area are free from obstacles.
The center of the building - the area between the three cores - is an open spaces illuminated by eliminating three floors of the office modules on each side. These large gaps become "windows" for the central spaces and allow for the free circulation of air. Externally, a skin of lattices with sliding panels becomes a sort of screen.

또레 꾸베

이 건물에서 우리는 과달라하라 시의 온화한 날씨를 최대한 활용하여 햇빛이 잘 들고 통풍이 잘되며 에어컨이 필요 없는 사무공간을 만들고 싶었다. 또한 부지가 번화한 업무 지구에 있기 때문에 건축주는 눈에 띄는 건물을 원했다.
건축면적은 4,800m²로 정해졌으나 건물의 높이는 제한이 없었다. 이에 모든 설비와 수직 동선이 포함된 세 개의 중앙 콘크리트 코어가 있는 고층건물을 짓기로 결정하였다. 이 코어는 건물의 유일한 지지 구조체이다. 다양한 크기의 거대한 캔틸레버 보가 콘크리트 슬라브를 지지하여 사무실과 주차장에는 시야를 가리는 장애물이 없다.
각 면마다 세 개 층을 제거하여 건물의 중심부로 빛이 들어오도록 하였다. 이 커다란 틈은 중앙 공간을 위한 '창문'이 되며, 공기가 자유롭게 순환되도록 한다. 외부에서는 미닫이 격자 외피가 일종의 차양막이 된다.

Carme Pinós

Torre Cube

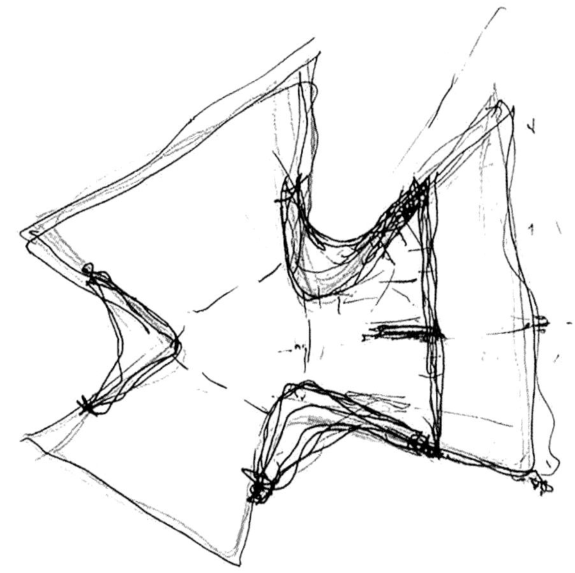

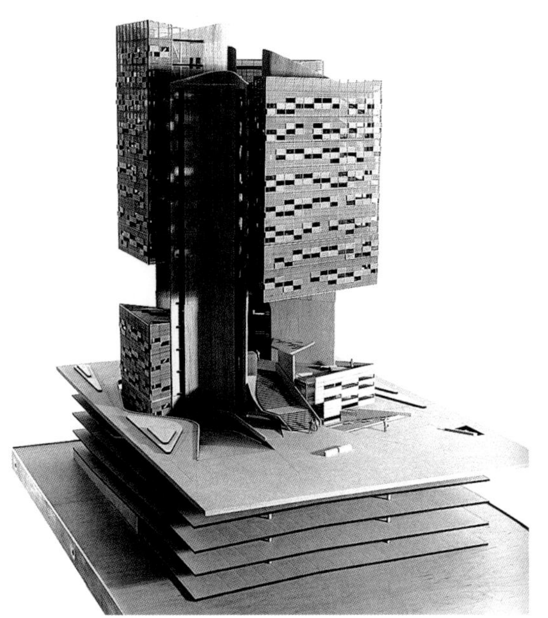

Architect Carme Pinós **Assistant architect** Juan Antonio Andreu, Samuel Arriola, Frederic Jordan, Cesar Vergés, Agustín Pérez, Holger Hennefarth, Caroline Lambrech **Structural engineer** Luis Bozzo **Construction Manager** Anteus **Client** Cube Internacional **Location** Guadalajara, Mexico **Design** 2002 **Completion** 2005 **Bldg. area** building - 7.000 m², parking - 10.000 m² **Cost** 10M USD **Photograph** Duccio Malagamba

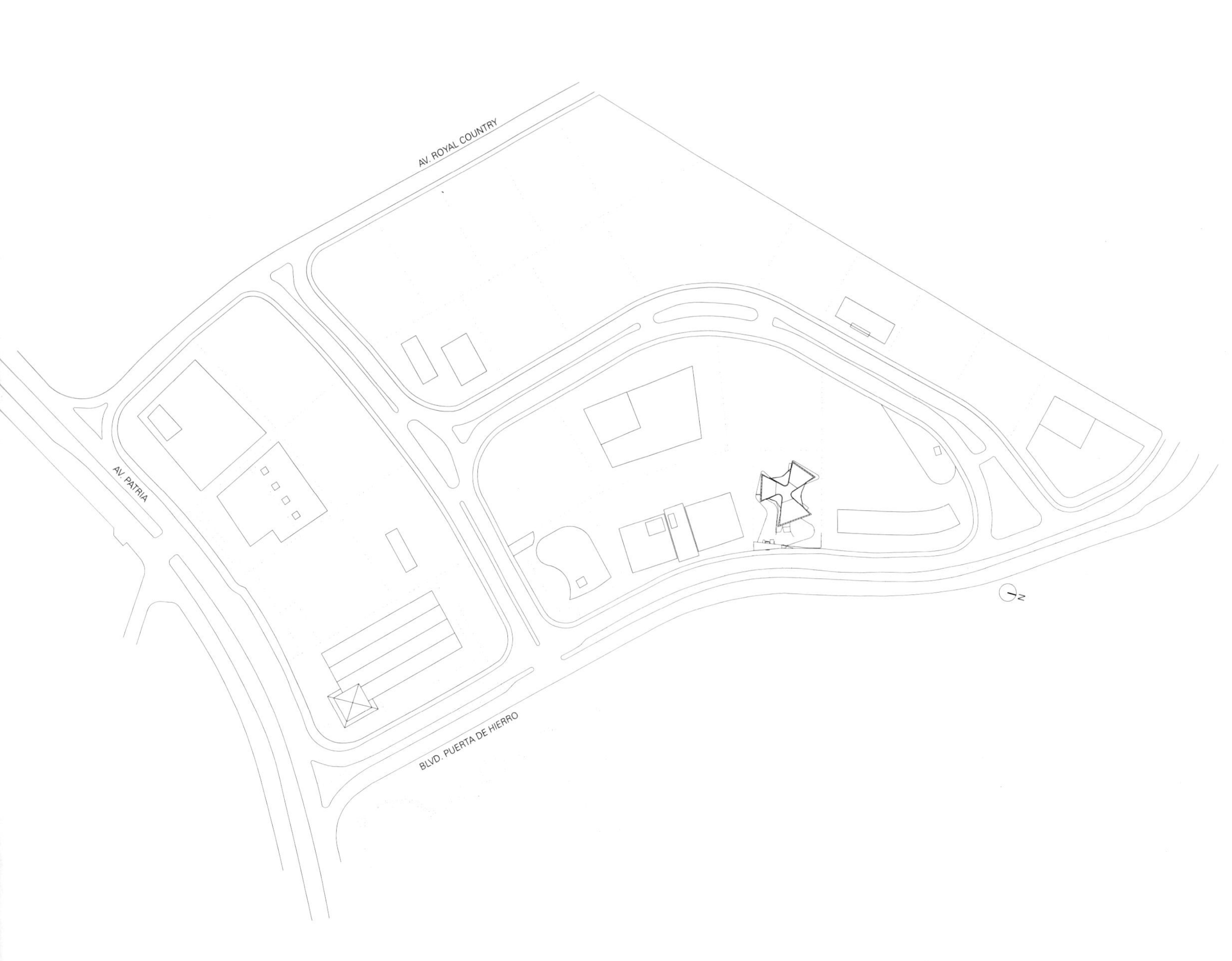
AV. ROYAL COUNTRY
AV. PATRIA
BLVD. PUERTA DE HIERRO
N

CUBE

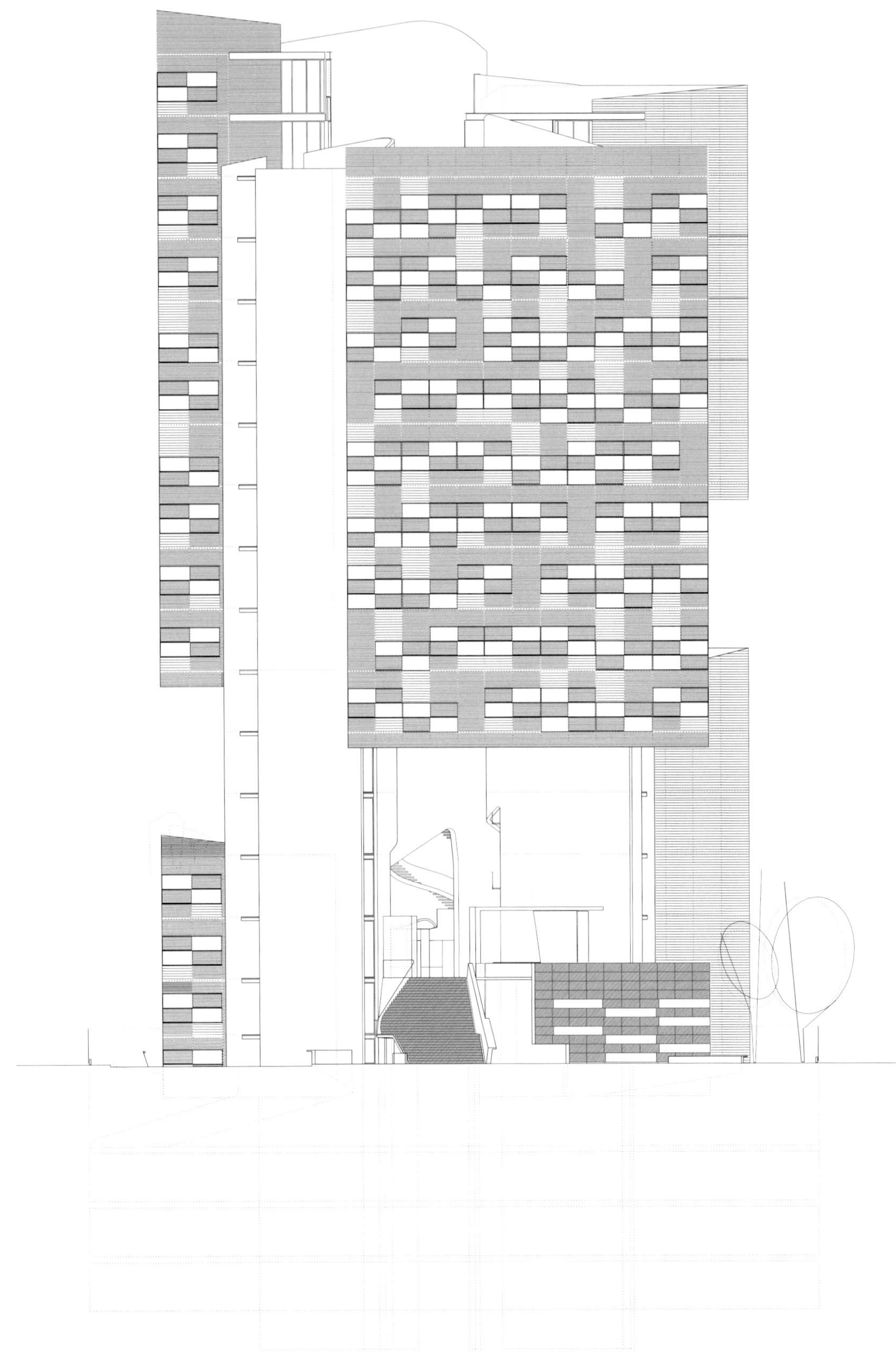

east elevation

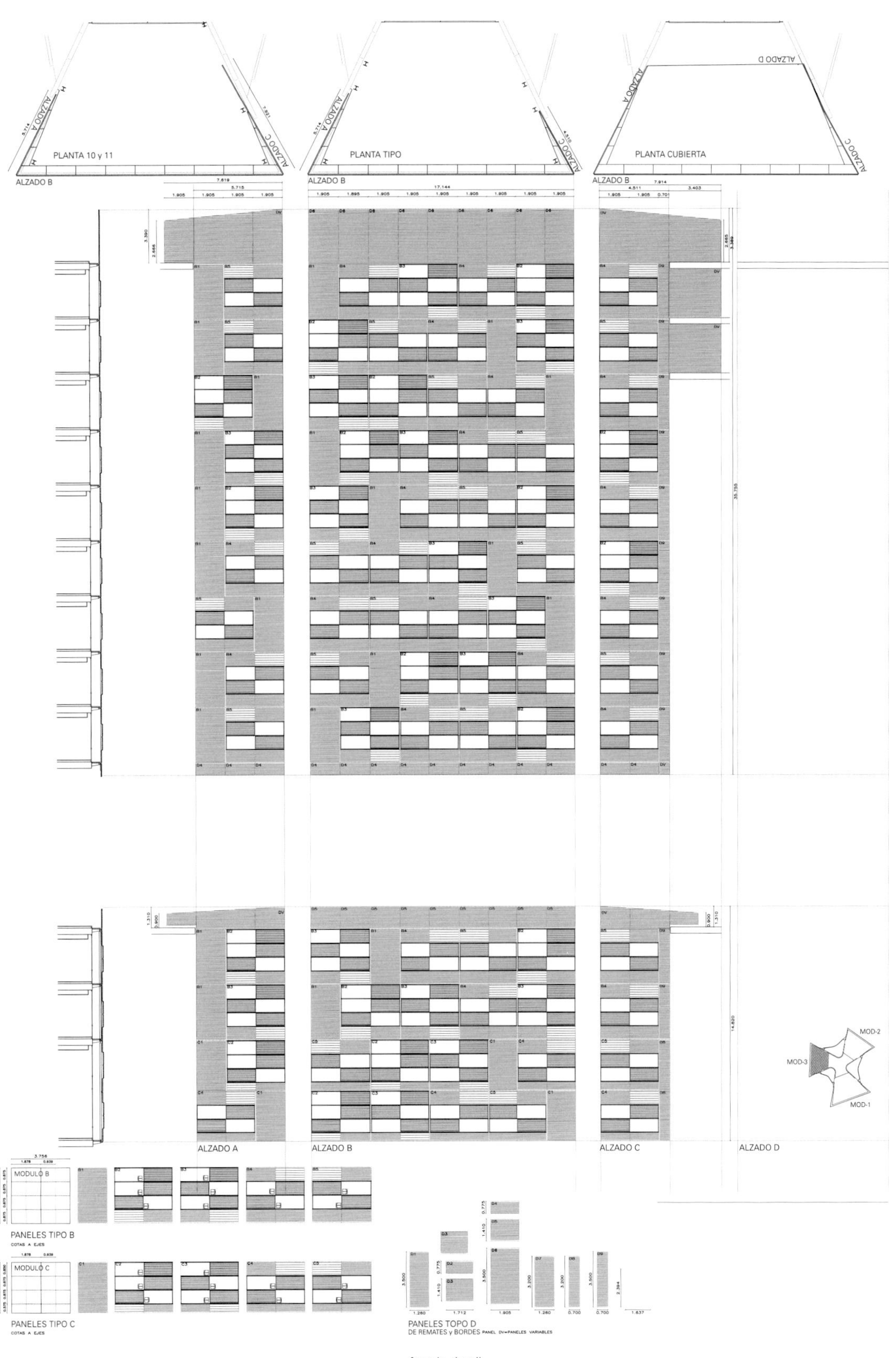

facade detail

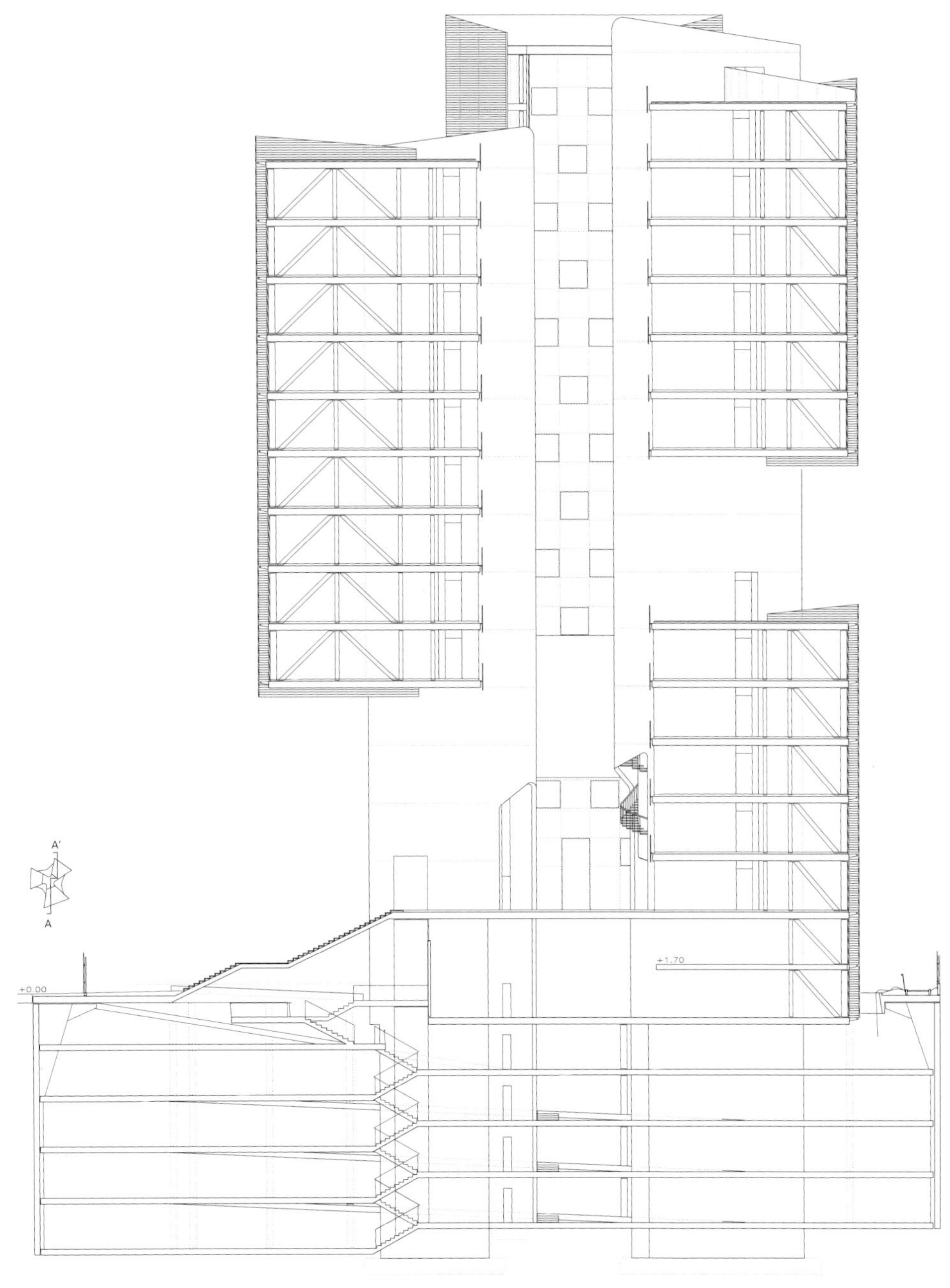

section A - A'

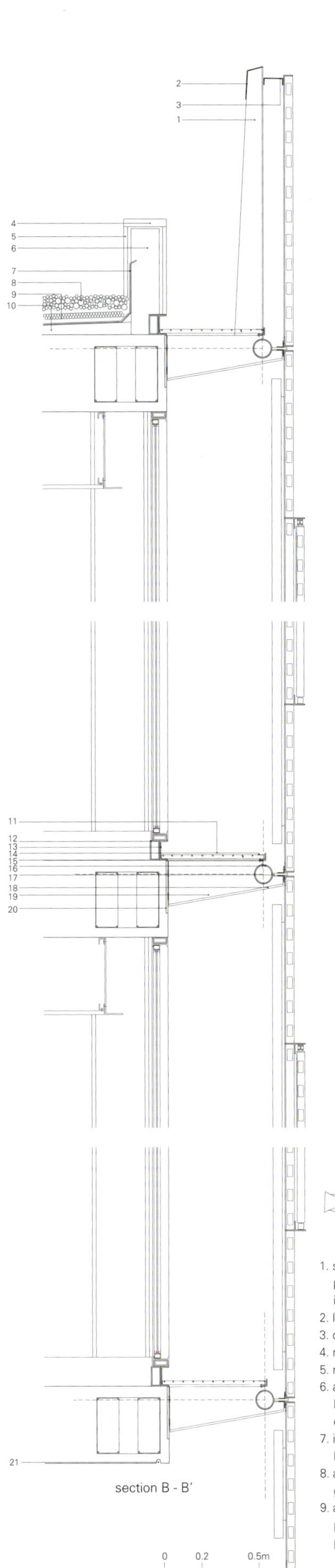

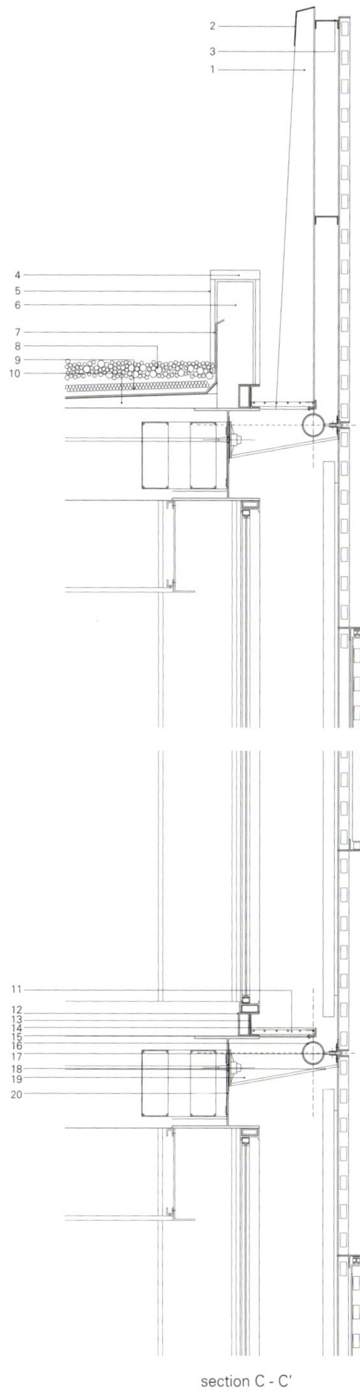

1. soporte paneles madera, formado por perfil metalico de acero inoxidable tipo T vatiable
2. lamina de acero e = 5mm
3. canal 101.2mm
4. recinto e = 40mm
5. recinto e = 20mm
6. antepecho formado por murete de ladrillo macizo, altura 45cm y espesor 15cm
7. impermeabilizacion formada por lamina asfaltica, solape 10cm
8. acabado superficial mediante grava ø20mm minimo
9. aislamiento termico formado por paneles machihembrados de poliuretano extruido, espesor 40mm
10. formacion de pendiente mediante hormigon con arlita, pendiente 2%
11. rejilla tipo irving marca metelmex galvanizada
12. perfil 102 x 51 mm
13. canal 101.6mm
14. fijacion de rejilla tipo mariposa marca metelmex galvanizada
15. angulo 4.8 x 38.1
16. tubo d = 102.9mm
17. solera 13 x 76 mm
18. patin cartela e = 10mm
19. alma cartela e = 5mm
20. placa de fijacion a losa 200 x 270 x 60 mm
21. goteron-separador losa hormigon perfil metalico ch. dimensiones 20 x 20 mm
22. cartela metálica formacion de banco
23. pavimento acabado

TORRE CUBE

first floor below ground

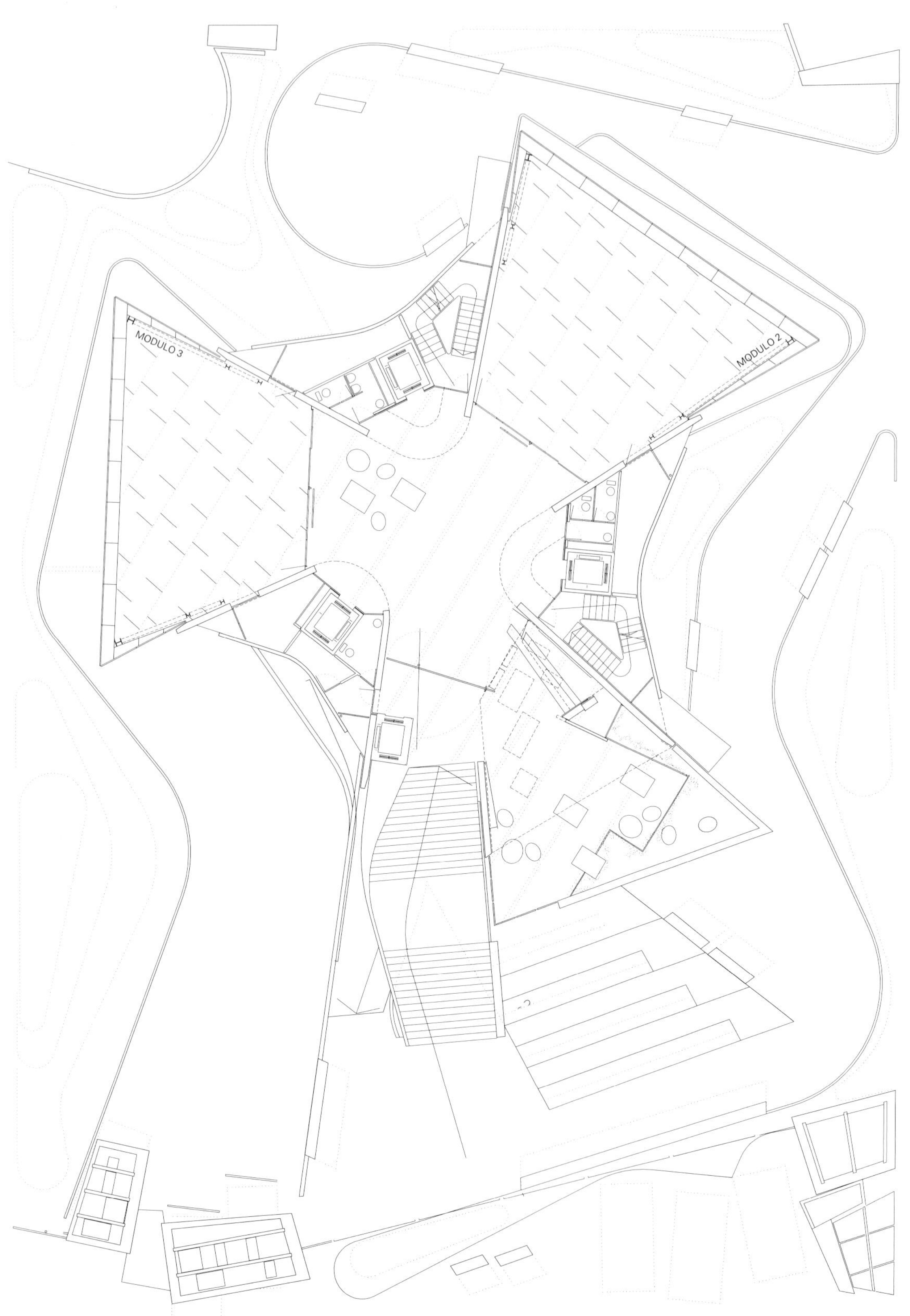

second floor (hall)

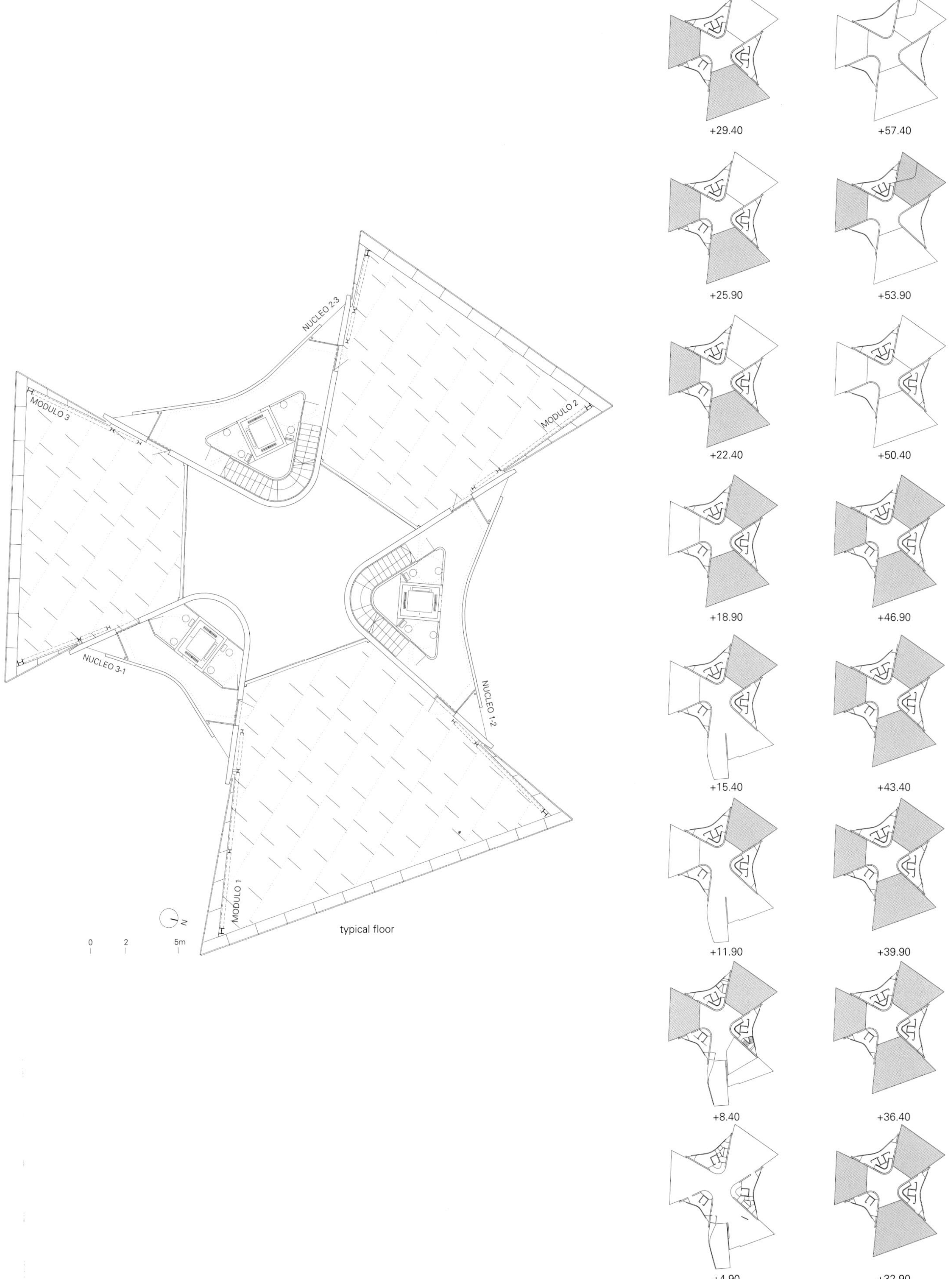

typical floor

The San Francisco Federal Building is a reflection of the GSA's commitment to design excellence and sustainable architecture.
As part of its commitment to sustainable living, the GSA works to reduce consumption of natural resources, minimize waste, and create a healthy and productive work environment for all tenants who occupy federal workspace. The San Francisco Federal Building is a demonstration of this commitment, incorporating state-of-the art technology and performance driven innovation. Tenants of the San Francisco Federal Building will enjoy sweeping views of San Francisco, and an abundance of natural light and natural ventilation that promote a healthy work environment while dramatically reducing energy consumption.
The new federal building is a slender 65-feet-wide tower rising 18 stories (240 feet) along the northern edge of the Mission and Seventh Street site. A four-story building annex adjoins the tower at the western edge of the site, helping to define the space that constitutes a new public plaza. In addition to this active plaza, the facility includes a number of resources that are available for public use, including a cafe, a childcare center, and a conference center.

Climate control

Throughout the year, San Francisco's design temperature ranges between 44 and 78 degrees Fahrenheit. The new San Francisco Federal Building takes advantage of the temperate climate to provide a comfortable interior environment while reducing energy consumption.
As a whole, the building is best understood as a hybrid that includes different space conditioning strategies appropriate for different locations in the building. The first five levels, with

미 연방정부 빌딩

샌프란시스코 연방정부 빌딩은 우수한 설계와 지속 가능한 건축에 대한 연방조달처GSA의 열망을 반영한다.
이를 위해 천연 자원의 소비를 줄이고, 쓰레기를 최소화하며, 연방정부에서 일하는 모든 근무자들에게 쾌적하고 생산적인 환경을 만들어주고자 하였다. 샌프란시스코 연방정부 빌딩은 이 노력의 증거이며, 혁신적인 기술과 실행이 통합된 건축이다. 이곳에서 일하는 근무자들은 샌프란시스코의 전경을 바라보는 행운을 누리게 될 것이며, 풍부한 자연광과 자연 환기로 에너지 소비는 최소화하면서 쾌적한 근무환경에서 일하게 될 것이다.
이 새로운 연방정부 빌딩은 65피트 너비의 날씬한 18층 높이 타워로 7번가와 미션 가의 북쪽 가장자리를 따라 서있다. 대지의 서쪽 면으로 타워와 연결된 4층 높이의 부속 건물은, 광장을 뚜렷이 나타내는데 한몫한다. 이 활기 넘치는 광장에는 식당, 탁아소, 강당을 포함하여 대중들이 사용할 수 있는 여러 가지 시설이 구비되어 있다.

실내 온도 조절

연중 샌프란시스코의 온도변화율은 화씨 44~78 정도이다. 이 빌딩은 온도기후를 이용하여 에너지 소비를 줄이는 동시에 안락한 내부 환경을 제공한다.

Morphosis

United States Federal Building

high concentrations of people and equipment, are fully air-conditioned. Above the fifth floor, the windows automatically adjust, allowing fresh air directly into the building for natural ventilation and free cooling. The window system creates a 'living skin' that allows the building to breathe. Breezes pass through openings on the windward side and are vented out through the leeward wall, with control based on wind speed and direction.

A computerized system, known as the building automated system (BAS), controls and monitors all of the building's mechanical equipment including those devices that are used to maintain internal environmental conditions and lighting levels. On the naturally ventilated floors, the computer system opens and closes windows, vents and sunscreens in response to temperature within the building as well as external environmental conditions. The window wall features manually operated windows for occupant control of the internal environment and includes a heating system integrated into the mullions. A minimal number of central, fully enclosed offices and meeting rooms are served by local, supplemental cooling units to accommodate higher density occupancies. During the night, the BAS opens the windows to flush out heat build-up and allows the nighttime air to cool the building's concrete interior. Throughout the day the thermal mass of the exposed concrete columns, shear walls and wave-form ceilings help cool the occupants of the building.

In the tower, the design of the high-performance facades is critical to the functioning of the natural ventilation. At the southeast elevation, a perforated metal sunscreen protects the glass facade from excess solar heat gain; at the northwest elevation, a series of fixed translu-

이 건물은 전체적으로 내부의 위치에 따라 적합한 냉난방 조절 방법을 적용한 혼합형 건물이다. 사람들과 설비가 가장 많이 모이는 하부 5개 층은 전체 냉방이 이루어지고, 6층부터는 창문이 자동적으로 조절되어 직접 들어오고 나가는 상쾌한 공기가 자연환기 및 냉각기능을 한다. 이 창문 시스템은 '살아있는 피부층'을 만들어 건물을 숨쉬게 하며, 산들바람이 환기창을 통해 들어와 반대편 벽으로 빠져나간다.

소위 빌딩 자동화 시스템[BAS]으로 알려져 있는 전산 시스템이 내부 환경이나 조도와 관련된 설비를 관리하고 감독한다. 자연환기를 이용하는 층에서는, 이 시스템이 건물 내부의 온도 및 외부 환경조건에 반응하여 자동으로 창문, 통기구, 차양을 열고 닫는다. 유리벽에는 사용자가 수동으로 내부 환경을 조절할 수 있는 창문이 달려 있으며, 커튼월 속에 난방 시스템이 통합되어 있다. 완전히 막힌 중앙의 일부 사무실 및 회의실은 부분적으로 냉방 유닛들을 갖추었다. 밤에는 전산 시스템에 의해 창문이 열려 축적된 열이 발산되고 밤공기를 건물로 유입시켜 내부 콘크리트의 온도를 낮춘다. 낮동안은 노출 콘크리트 기둥, 전단벽, 파형 천장과 같은 축열체가 건물을 선선하게 유지해준다.

고성능으로 설계된 타워의 입면은 자연통풍에 결정적 기능을 한다. 남쪽 입면의 펀칭 메탈 차양은 유리 입면이 과도한 태양열을 받지 않도록 막아주며, 북쪽 입면에는 반투명 고정 차양이 외부 통로에 장착되어 태양빛을 막고 그늘을 만들어준

Architect Morphosis - Thom Mayne **Executive architect** Smith Group, Inc. **Construction management** Hunt Construction Group **General contractor** Dick Corporation/Morganti General Contractors **Structural/Mechanical/Electrica/Plumbing engineer** Ove Arup and Partners **Natural ventilation modeling** Lawrence Berkeley National Laboratory **Landscape architect** Richard Haag Associates Inc. with J.J.R **Civil engineer** Brian Kangas Foulk **Geotechnical** Geomatrix **Lighting consultant** Horton Lees Brogden Lighting Design, Inc. **Signage** Kate Keating Associates **Cost estimator** Davis Langdon **Curtain wall** Curtain Wall Design & Consulting, Inc. **Blast consultant** Hinman Consulting Engineers **Code** Rolf Jensen & Associates **Acoustics** Thorburn Associates **Vertical transportation** Hesselberg, Keessee & Associates, Inc. **Client** U.S. General Services Administration **Location** 7th and Mission Streets, San Francisco, California, USA **Site area** 91,000 ft^2 **Gross floor area** 605,000 ft^2 **Design start** 2000 **Construction** 2003~2007 **Completion** 2007 **Cost** 144M USD **Photograph** Roland Halbe

cent sunshades are attached to an exterior catwalk, breaking the sun's path to shade the glass These climate specific facades give the building its distinctive appearance.

Nationally, the GSA strives to use no more than 55,000 Btu of energy per square foot per year in its buildings. The new San Francisco Federal Building is expected to surpass the GSA's target as well as California's stringent Title 24 Energy Code. The naturally ventilated floors are projected to have an average energy consumption of less than 25,000 Btu per square foot per year, a significant improvement over the national standard. The local utility provider, PG & E, has confirmed that the project will receive an energy rebate of $250,000 at the completion of construction.

Electrical

Lighting is typically the largest energy cost for an office building, representing up to 40 percent of a facility's total energy load. The new San Francisco Federal Building's lighting strategies improve the workplace and are a critical facet of this project's sustainable design. Approximately 85 percent of the workspace is illuminated with natural light.

Ambient light, the general illumination in an office, comes from sunlight channeled through the windows and reflected off walls and ceilings to extend its reach with minimum glare and intensity. With an average overall ceiling height in the tower of 13 feet, natural daylight will penetrate deep into work spaces. Powered lights are also provided to supplement the natural light. Through simple sensors, the building's automated systems manage the balance between powered and natural daylight. The powered lights are on only when people are at their worksta-

다. 기후에 따른 이러한 입면 장치로 인해 건물은 독특한 외형을 갖게 되었다.

이 건물은 평방피트 당 연간 에너지 소비량이 55,000 BTU를 넘지 않도록 하는 GSA의 목표뿐 아니라 캘리포니아 주의 엄격한 에너지 규정도 수월하게 넘길 수 있을 것으로 기대된다. 자연 환기 층은 연간 평균 에너지 소비량을 평방피트 당 25,000 BTU 이하를 목표로 한다.

전기설비

사무실 건물에서 일반적으로 에너지 비용을 가장 많이 차지하는 것은 조명이다. 조명은 보통 전체 에너지 부하의 40%까지도 차지한다. 근무 공간을 향상시키며 지속 가능한 환경을 위해 이 빌딩은 근무지의 약 85%를 자연광으로 조절한다.

사무실의 전반적인 채광은 창문을 통해 들어오는 햇빛을 벽과 천장으로 반사시켜 눈부심과 빛의 세기를 최소화하는 동시에 빛이 뻗어 들어오는 범위를 넓혀준다.

타워의 평균 층고는 13피트로, 이는 자연광이 사무실 내부 깊숙이 들어올 수 있는 높이이다. 자연광을 보완하기 위해 인공 조명도 제공된다. 센서를 이용한 자동 시스템이 조명과 자연광 사이의 균형을 조절한다. 인공 조명은 사무실에 사람이 있을 때만 작동한다. 이 혼합 조명으로 조명에 쓰이는 에너지를 약 26% 줄일 수 있다.

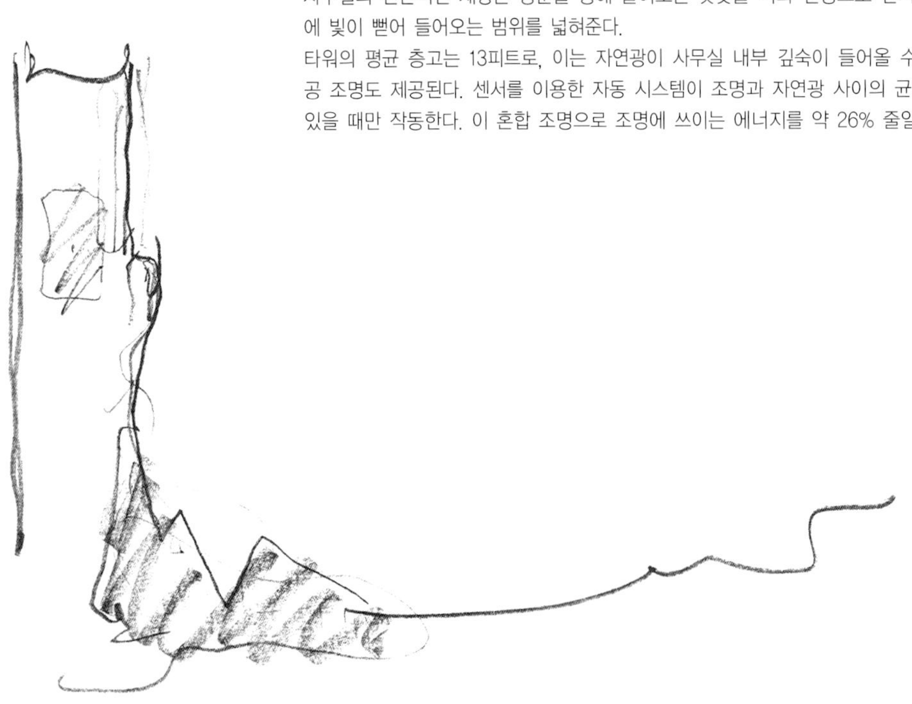

tions. Together, these approaches reduce energy used for lighting by approximately 26 percent.

Flexible information technology

Large and flexible floor plans with power and data cabling conveniently routed through easily accessible, under floor space make workspace reconfiguration easy while reducing costs for remodeling.

Building materials

The San Francisco Federal Building incorporates building materials and construction strategies that minimize waste and energy consumption.

The building minimizes pollution by replacing high proportions of Portland cement in its concrete foundations and frame. During the manufacturing process, Portland cement is associated with very high levels of greenhouse gas emissions. In the Federal Building°'s concrete mixture, 50% of the pollution-intensive Portland cement is replaced with blast furnace slag, a recycled waste product from the steel industry, significantly reducing greenhouse gas emissions resulting from conventional concrete. This environmentally sound choice also results in higher-strength concrete and has a warm, light-colored tone that contributes to the favorable daylight penetration within the office space.

The GSA mandated that 75 percent of materials used during construction be recycled. Currently, the project is recycling 87 percent of its waste material.

Carpet, paint and furniture were carefully considered with respect to the project's sustainable goals.

탄력적 정보기술

바닥은 동력 및 데이타 전선과 함께 크고 탄력적으로 설계되어, 쉽게 접근할 수 있는 바닥 아래에 깔려있다. 바닥 아래의 설비(액세스플로어)로 리모델링 시 부품교체가 쉬운 동시에 비용도 절약된다.

건축 자재

이 빌딩은 폐기물과 에너지 소비를 최소화할 수 있는 건축자재와 건설 전략을 적용하였다.

건물은 콘크리트 기초 및 골조에 높은 비중을 차지하는 포틀랜드 시멘트를 다른 자재로 대체하여 오염을 최소화하였다. 포틀랜스 시멘트는 제조하는 과정에서 배출되는 이산화탄소량이 매우 높기 때문에 전체량의 50%를 철강 산업에서 나오는 폐기물을 재활용하여 만든 고로 슬래그 골재로 대체하여, 이산화탄소 방출량을 현저히 줄였다. 이러한 친환경적 선택으로 고강도의 콘크리트 시공은 물론 밝고 따뜻한 색감으로 사무실 내부를 안락한 빛으로 채울 수 있게 되었다.

또한 이 건물은 건설에 사용된 자재의 75%을 재활용하였다. 실재로, 이 프로젝트에 사용되고 남은 폐자재의 87%가 재활용되고 있다. 카페트, 페인트, 가구 등도 지속가능성을 고려하여 정하였다.

SAN FRANCISCO FEDERAL BUILDING

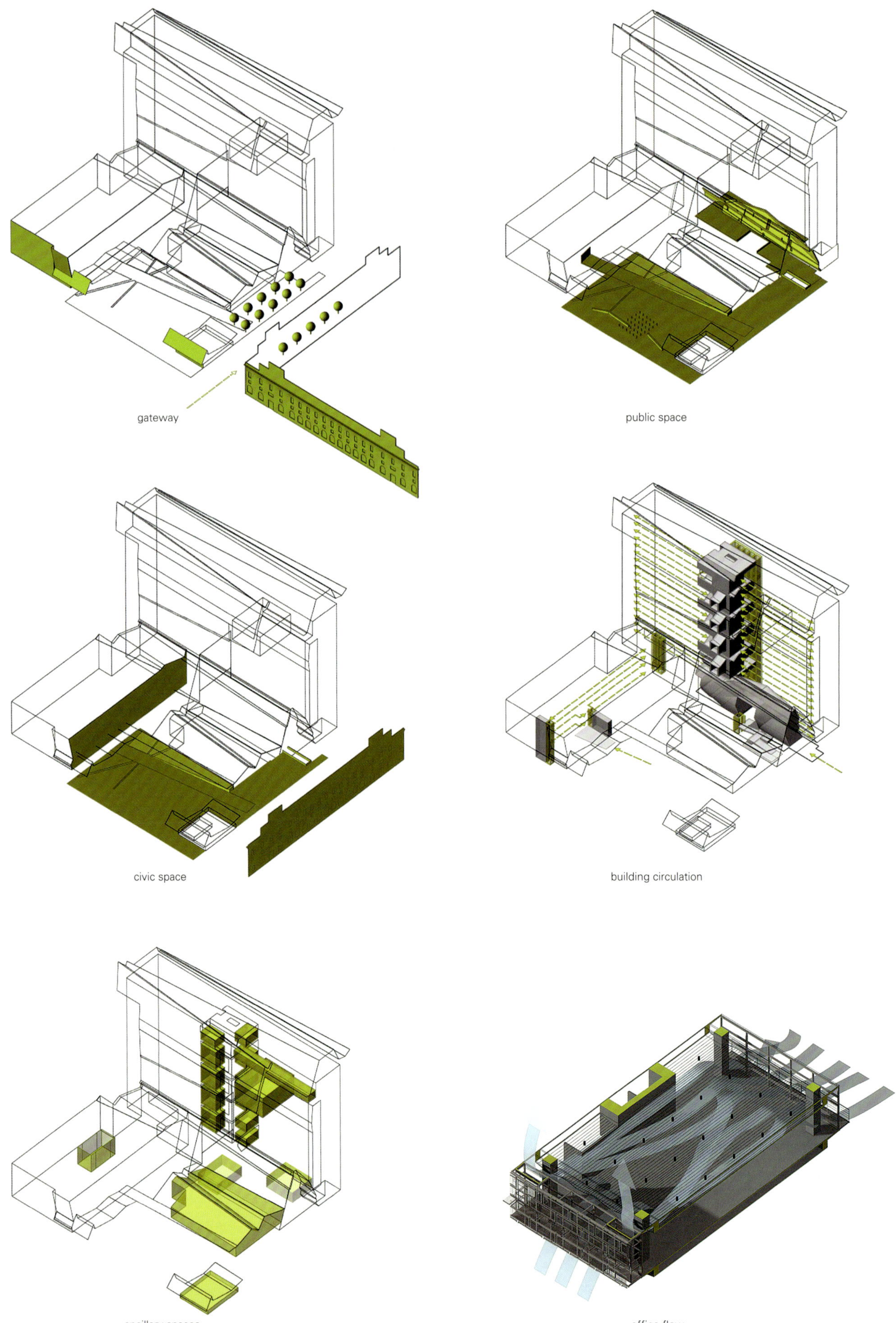
gateway
public space
civic space
building circulation
ancillary spaces
office flow

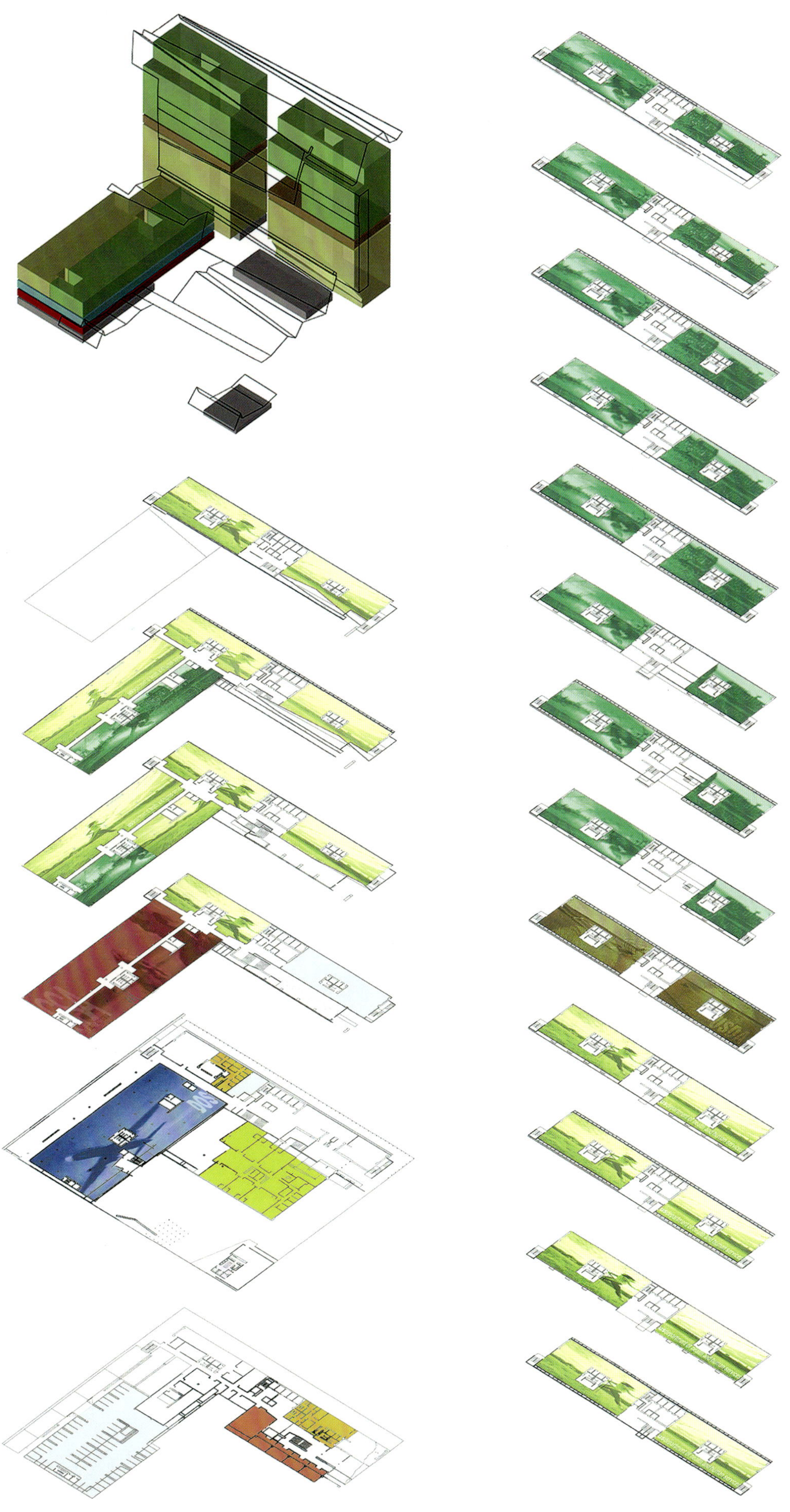

SAN FRANCISCO FEDERAL BUILDING

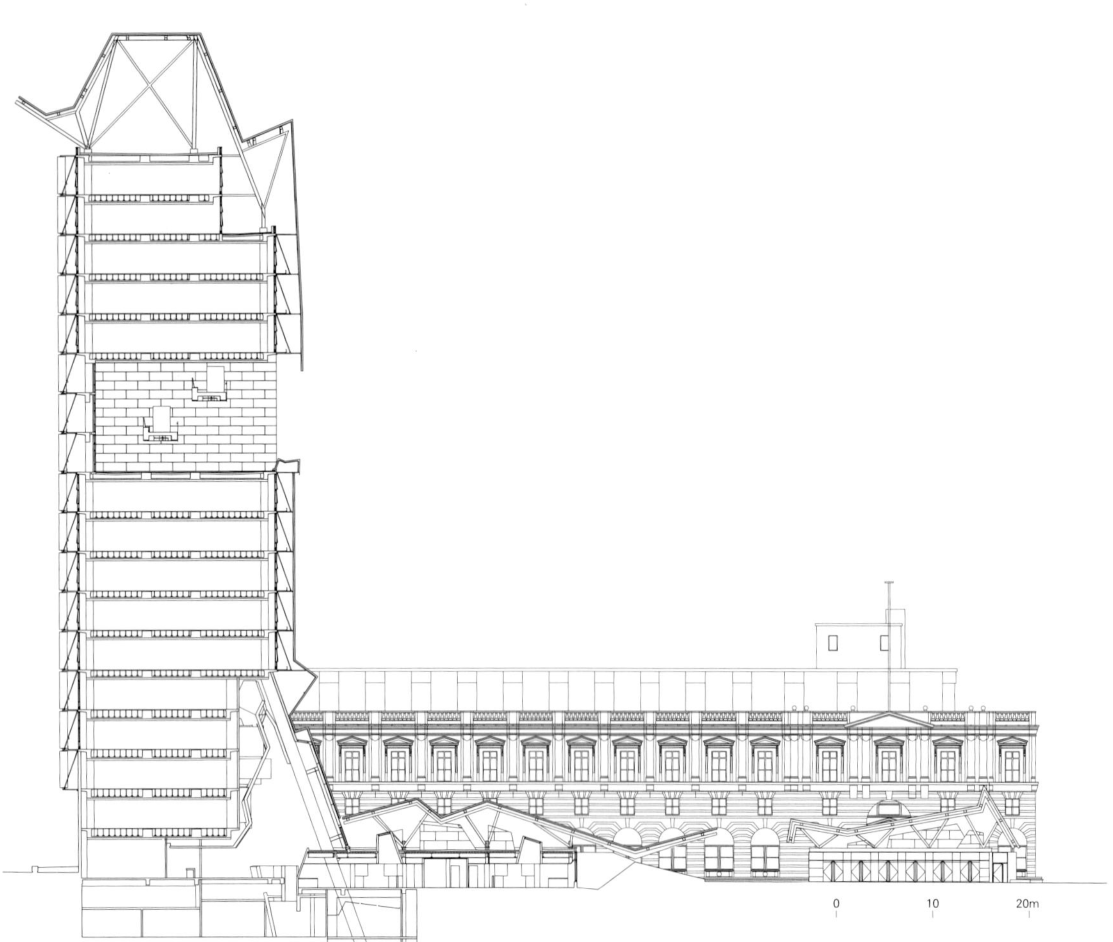

section through sky garden

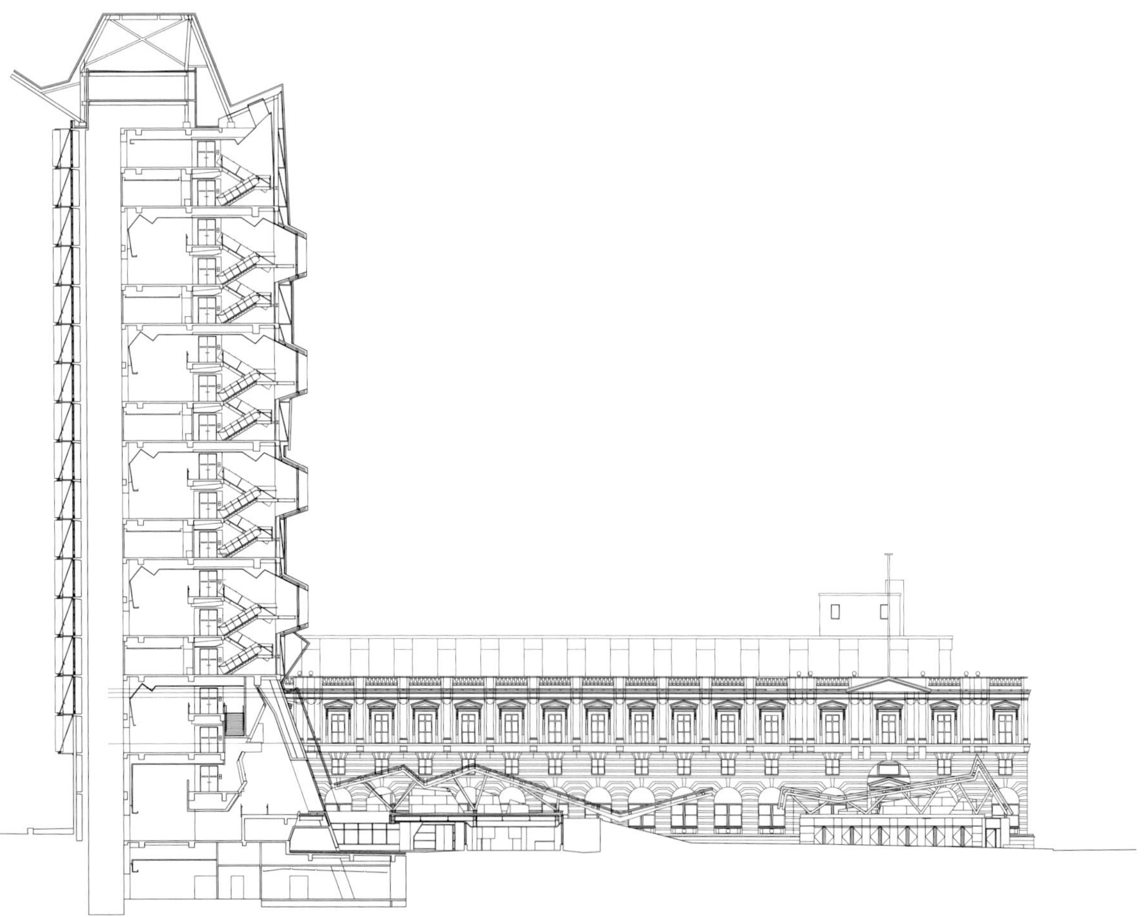

section through elevator lobby

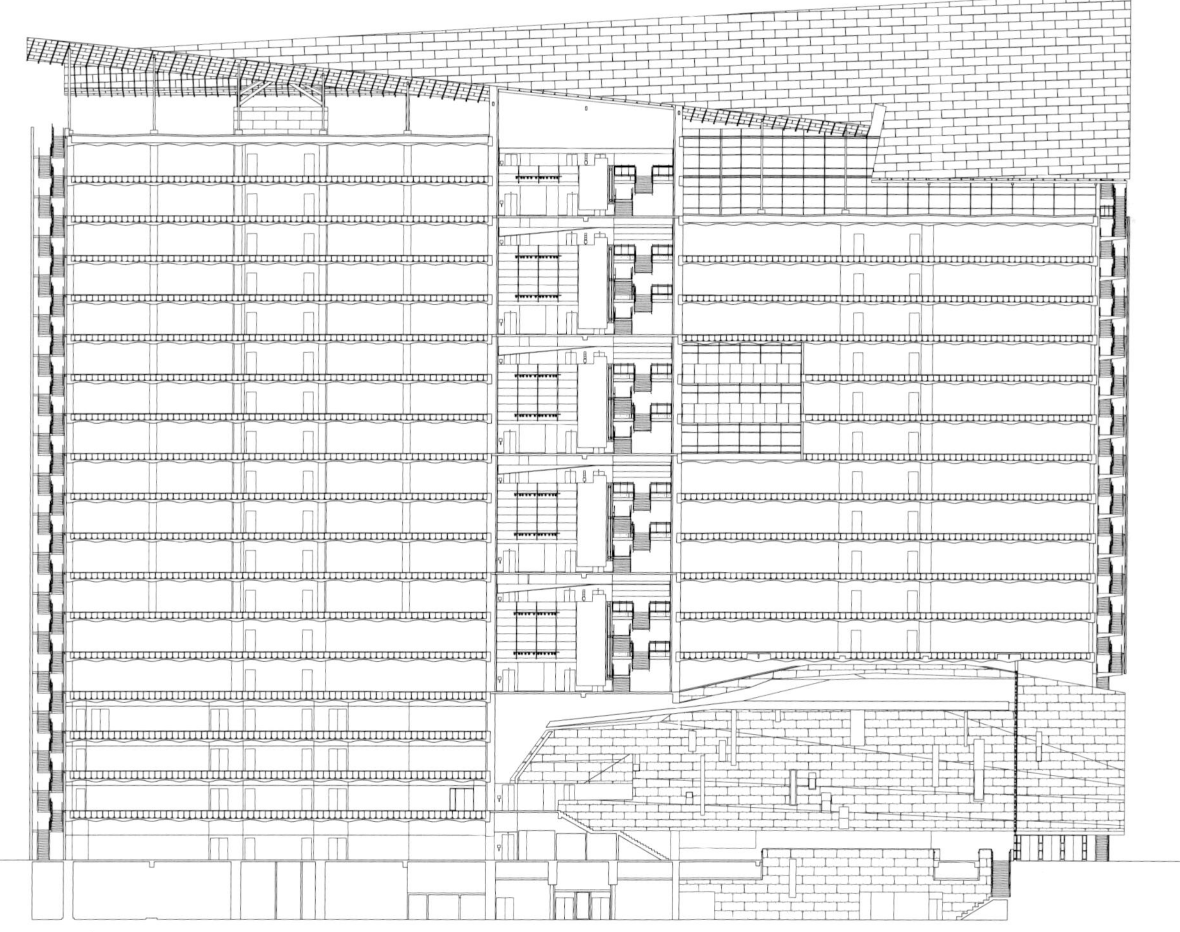

longitudinal section

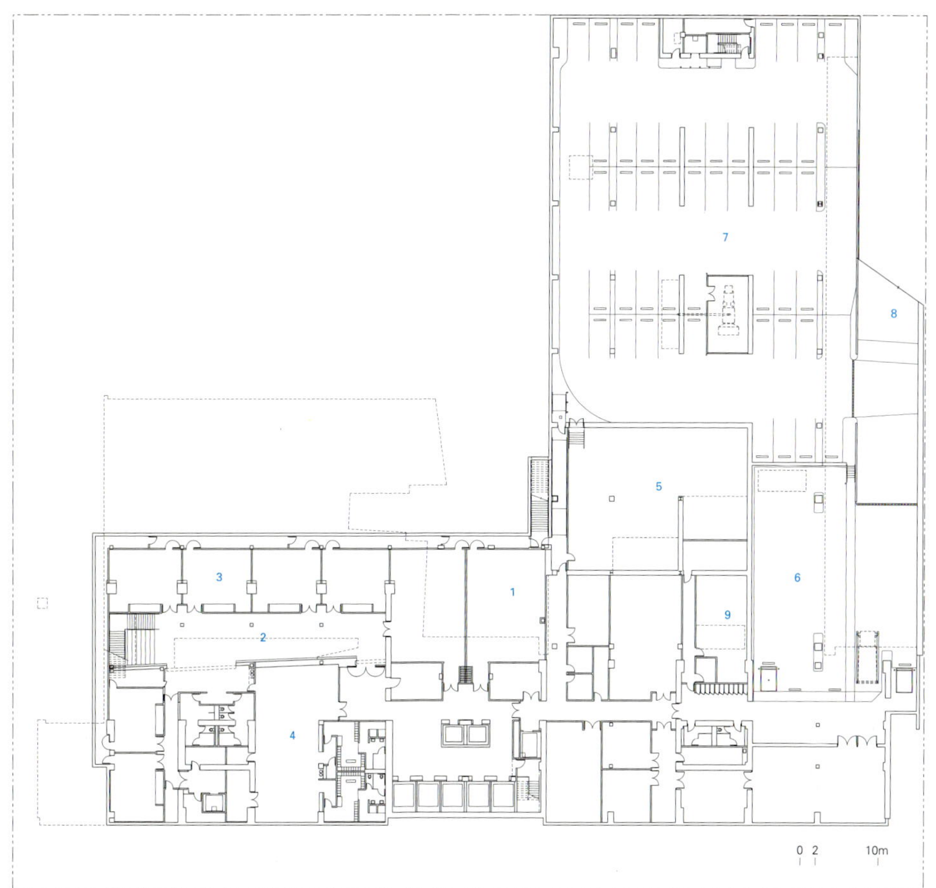

basement

8
11
13
12
14
10
2

first floor

1. auditorium
2. lobby
3. conference facility
4. fitness center
5. service
6. loading
7. parking
8. auto ramp access
9. mailroom
10. tower entry
11. plaza
12. annex entrance
13. cafe pavilion
14. daycare

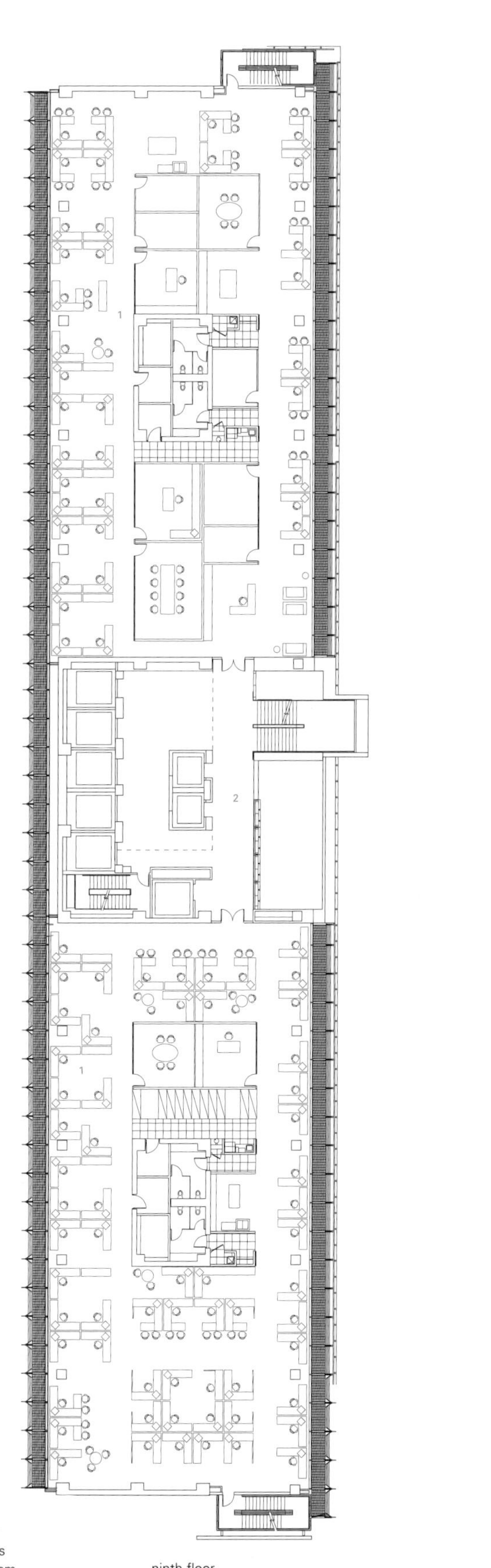

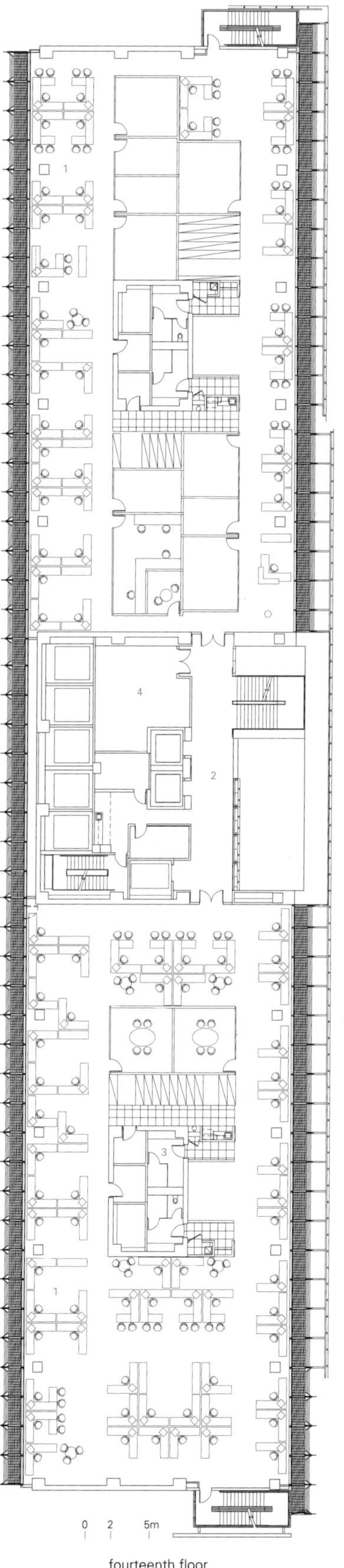

1. office space
2. elevator lobby
3. restroom cores
4. conference room

ninth floor

fourteenth floor

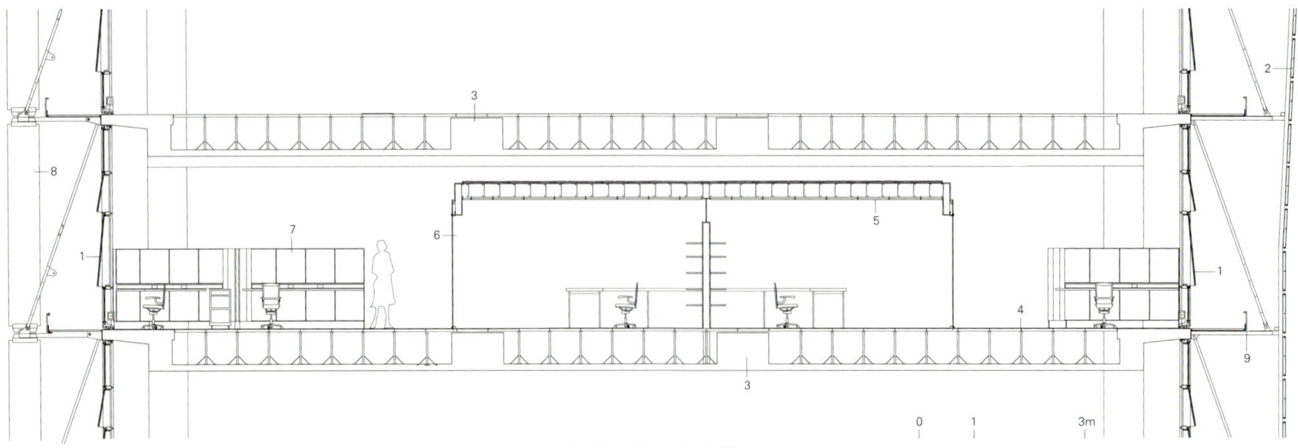

section through a typical office

1. metal/glass curtain wall
2. perforated stainless steel sunscreen panels
3. tower: concrete sheer wall system
4. raised flooring
5. acoustical ceilings
6. office partitions
7. office furniture: Herman Miller
8. laminated glass fin sunshade system
9. scrim support

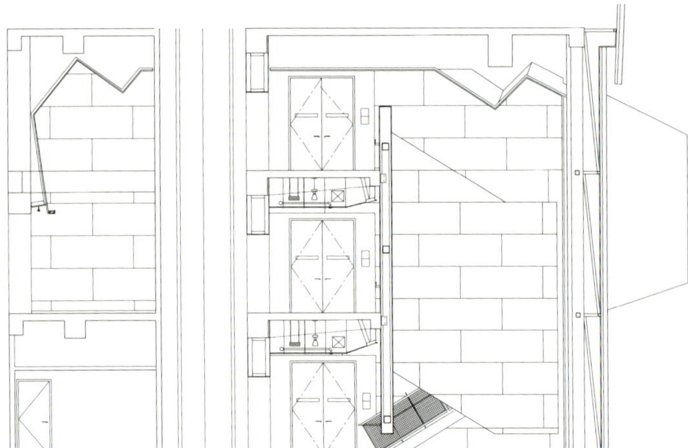

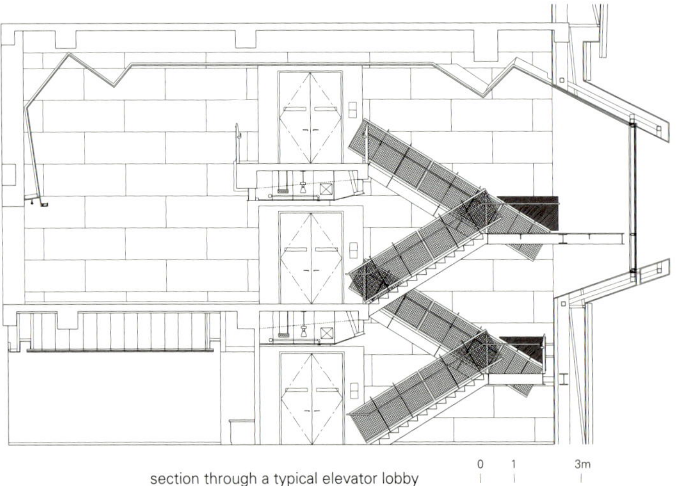

section through a typical elevator lobby

MAPS, DNA AND SPAM

The brief for this project was to create a large covered space that protects approximately 9000 pilgrims from rain and bad weather during the winter months and accommodates other, non-religious activities, while not disturbing the continuity of the historical place.
The new Church has been located between the historic Basilica and the Pastoral Centre, which facilitates easy access for large numbers of people. It creates an axis of religious activities as well as visual contact with these other focal points.
Basic principles of the design encompass a sensitive approach to the site, interaction with the existing elements of religious activities and an architecture that uses simple and yet grand means to evoke the desired religious effect, while at the same time being a fully functional multi-purpose hall.
Three basic elements, a rectangle, a circle and a linear element intersecting these, are the geometric elements that dominate the architecture.
The whole complex is developed semi-underground so that the apparent volume of the Church is reduced, gently rising from the existing square level. Three underground chapels in front of the cylindrical volume of the Church have been designed together with rooms for confessions, all of which are naturally lit through openings in the roof above.
The main structure is circular and despite its size embraces the pilgrims generously while maintaining the feeling of a Church, giving a panoramic view of the interior space and provid-

삼위일체 수도회 성당

이 프로젝트의 목적은 이 장소의 역사성을 해치지 않으면서 겨우내 이곳을 방문하는 9천여 명의 성지 순례자들을 악천후로부터 보호하고, 비종교적 활동까지 지원할 수 있는 대규모 공간을 짓는 것이었다.
이 교회는 역사 깊은 카톨릭 대성당과 목회 센터 사이에 있기 때문에 많은 사람들이 쉽게 접근할 수 있으며, 두 중심부가 시각적으로 만나게 하고 동시에 종교 활동 축을 생성한다.
설계의 기본 방침은 부지에 다한 섬세한 이해, 기존의 종교 건물과의 상호 작용, 그리고 단순하지만 높은 종교적 효과를 거둘 수 있으면서, 충분히 기능적인 다목적 홀을 짓는 것이었다.
직사각형, 원, 그리고 이 두 도형을 가로지르는 직선, 이 세 가지가 이 건축물의 주된 기하학 요소이다.
건물 전체는 반 지하로 계획되어 교회 볼륨은 외관상 축소되어 있으며, 광장 레벨에서부터 완만하게 올라온다. 교회의 원통형 볼륨 앞쪽에 있는 세 개의 지하 예배당은 고해실과 함께 설계하였다. 모든 예배당은 지붕을 통해 자연 채광된다.

Alexandros N. Tombazis and Associates Architects

Church of the Most Holy Trinity

Sanctuary of Fatima, Portugal

ing the desired monumentality. The walls of the building are clad in sandstone and have minimal openings. The roof structure is designed so as to provide excellent daylighting through north-facing clerestory windows as well as an integrated daylighting/artificial lighting solution that allows a wide variety of lighting scenarios for a great variety of purposes. The roof is intersected by two concrete beams, which rise towards the altar in a counter movement to the down sloping of the floor. It emphasizes the vertical focus related to the purpose of the place and incorporates south-facing natural lighting and translucent panels that emphasize the procession from the entrance to the presbytery and allow a variety of lighting scenarios.
The daylighting strategy, as part of an overall holistic bioclimatic design strategy, reduces the building's energy consumption, running costs and provides good comfort conditions for the users.
The interior space is clad in simple materials that support the monolithic and contemplative character of the structure. Walls and the central beams are clad in acoustic absorbing plaster, wood and marble.
All in all, the developed complex is aimed at providing a synthesis of the unique historic place of pilgrimage and the present-day requirements for a Church of great importance in a way that enhances the old, the new and the eternal.

기본 구조는 원형으로, 거대하지만 교회라는 느낌을 유지하며, 내부 공간의 전경과 기념비적 의미를 충분히 전달하면서 성지 순례자들을 관대하게 포용한다. 건물의 벽에는 사암을 입혔으며 창문은 최소화하였다. 지붕 구조는 북향의 채광층에 달린 창문을 통해 직사광을 충분히 받도록 설계하였다. 또한 다양한 목적에 맞는 조명이 가능하도록 자연광 및 인공 조명의 계획도 통합적으로 이루어졌다. 지붕을 가로지르는 두 개의 콘크리트 보는 위로 제단을 향하고 아래로는 바닥으로 기울어져 있다. 이것은 장소의 목적에 맞도록 중심의 수직성을 강조하며, 남쪽의 자연 채광창과 반투명 패널을 통합시켜 준다. 또한 입구부터 사제석을 향한 행렬을 강조하며 다양한 조명 계획을 가능하게 한다.
전체 생태기후계획에서 자연광은 건물의 에너지 소비, 유지비를 줄여주는 동시에 사용자에게 쾌적하고 편리한 환경을 조성한다. 내부 공간은 구조의 단일성과 묵상적 성격에 걸맞게 단순한 자재를 선정하였다. 이에 벽과 중앙 보는 흡음 석고 플라스터, 목재, 대리석으로 마감하였다. 결론적으로, 완성된 건물은 성지순례의 특별한 역사적 장소로서 현대 교회가 요구하는 구약, 신약, 그리고 불변성의 통합적인 제공을 목표로 한다.

Architect Alexandros N. Tombazis **Design team** Alexandros N. Tombazis and Associates Architects - Stavros Gyftopoulos, Sophia Paraskevopoulou **Local architect** Paula Santos Arquitectos, Lda. - Paula Santos, Joana Delgado **Structural design, security and hygiene planning** Eteclda - Escritório Técnico de Engenharia Civil, Lda. **Mechanical engineers** Edifícios Saudáveis - Consultoria, Lda. **Sanitary and sewage engineers** Vitor Abrantes -Consultoria e Projectos de Engenharia, Lda. **Electrical engineer** OHM - E, Gabinete de Engenharia Electrotécnica, Lda. **Acoustics and electroacoustics design** Vitor Abrantes - Consultoria e Projectos de Engenharia, SOPSEC, Lda., LPL, Lda. **Landscape design** Laura Costa **Energy consultants** University of Athens, Department of Applied Physics - Prof. M. Santamouris **Natural & artificial lighting** Bartenbach Lichtlabor GmbH **Acoustics consultant** C.S.T.B. **Graphic design** Francisco Providência **Client** Sanctuary of Fatima **Advisor** E. Corsepius **Location** Fatima, Portugal **Photograph** Courtesy of the architect ©FG+SG Fotografia de Arquitectura, Alexandros N. Tombazis - p.256, 262

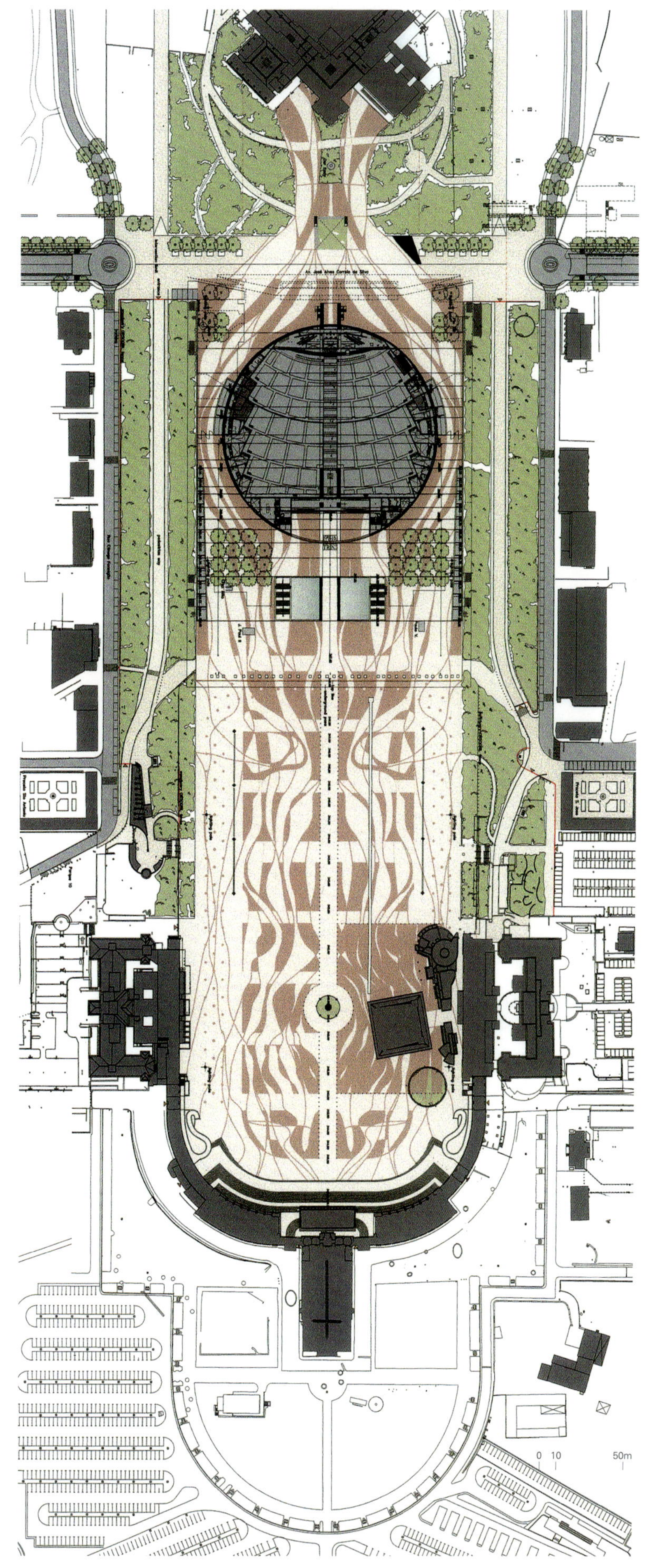
0 10
50m

• one space connecting people
proximity to altar
one focal point of attention

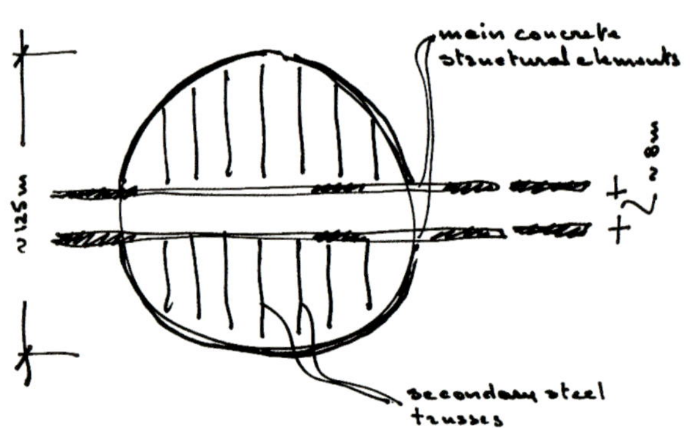

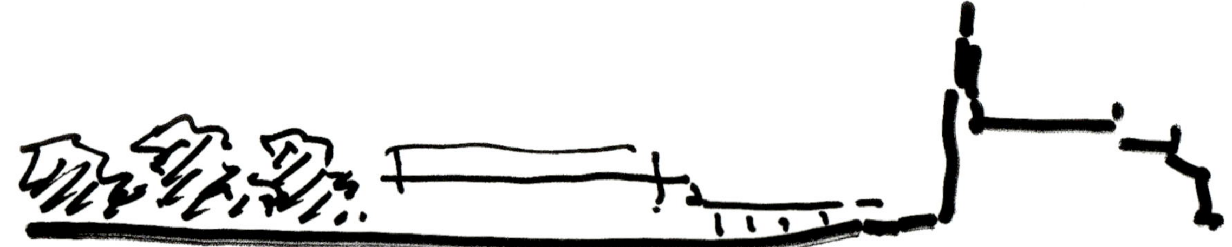

the importance of the longitudinal section

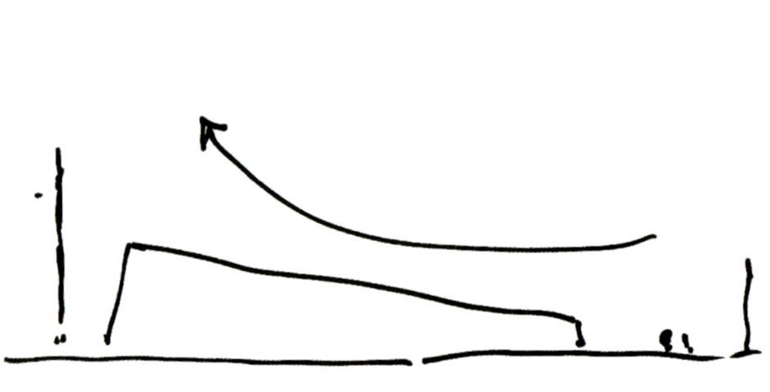

an up lifting of spirit and mind

11.10.07

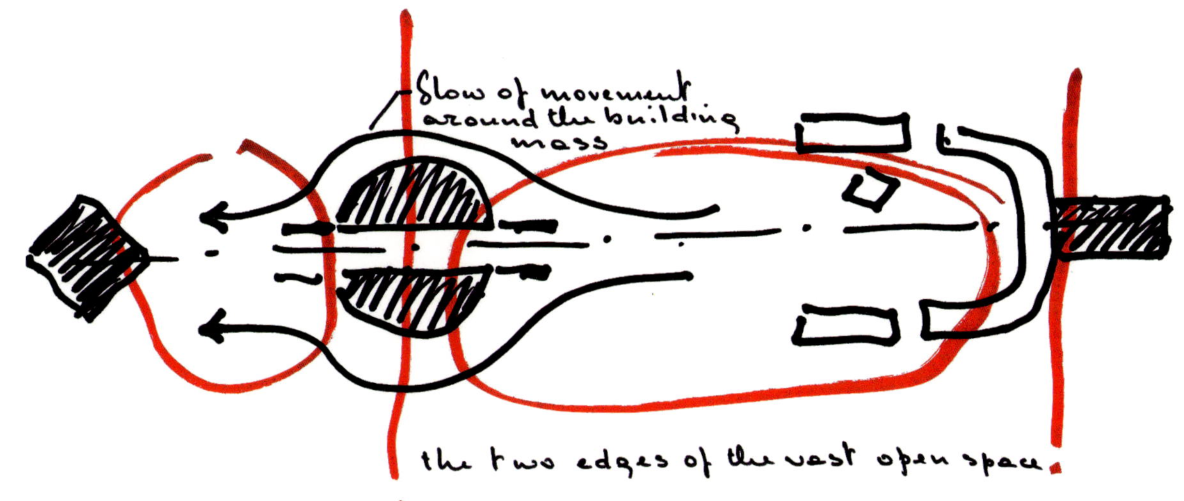
flow of movement around the building mass
the two edges of the vast open space

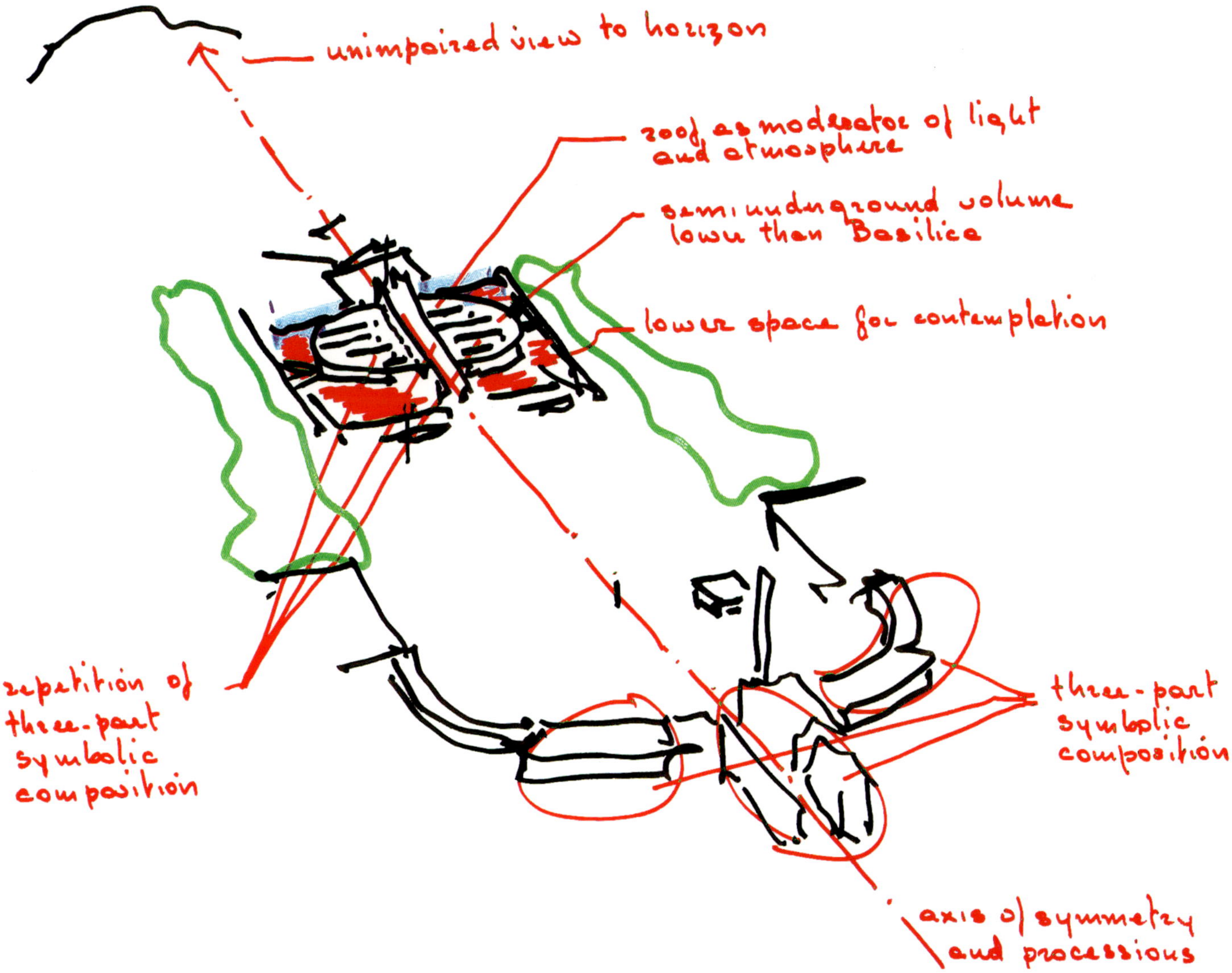

unimpaired view to horizon
roof as moderator of light and atmosphere
semi underground volume lower than Basilica
lower space for contemplation
repetition of three-part symbolic composition
three-part symbolic composition
axis of symmetry and processions

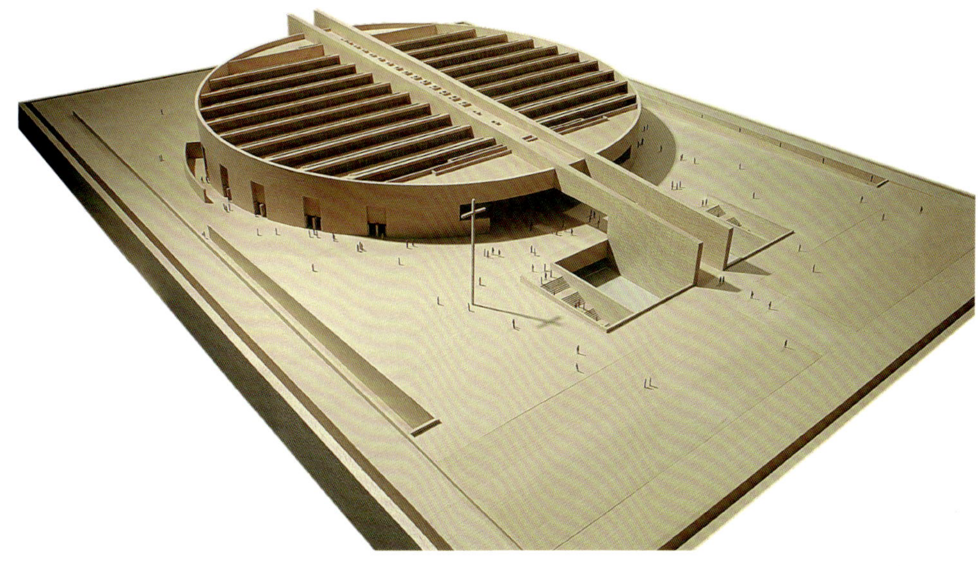

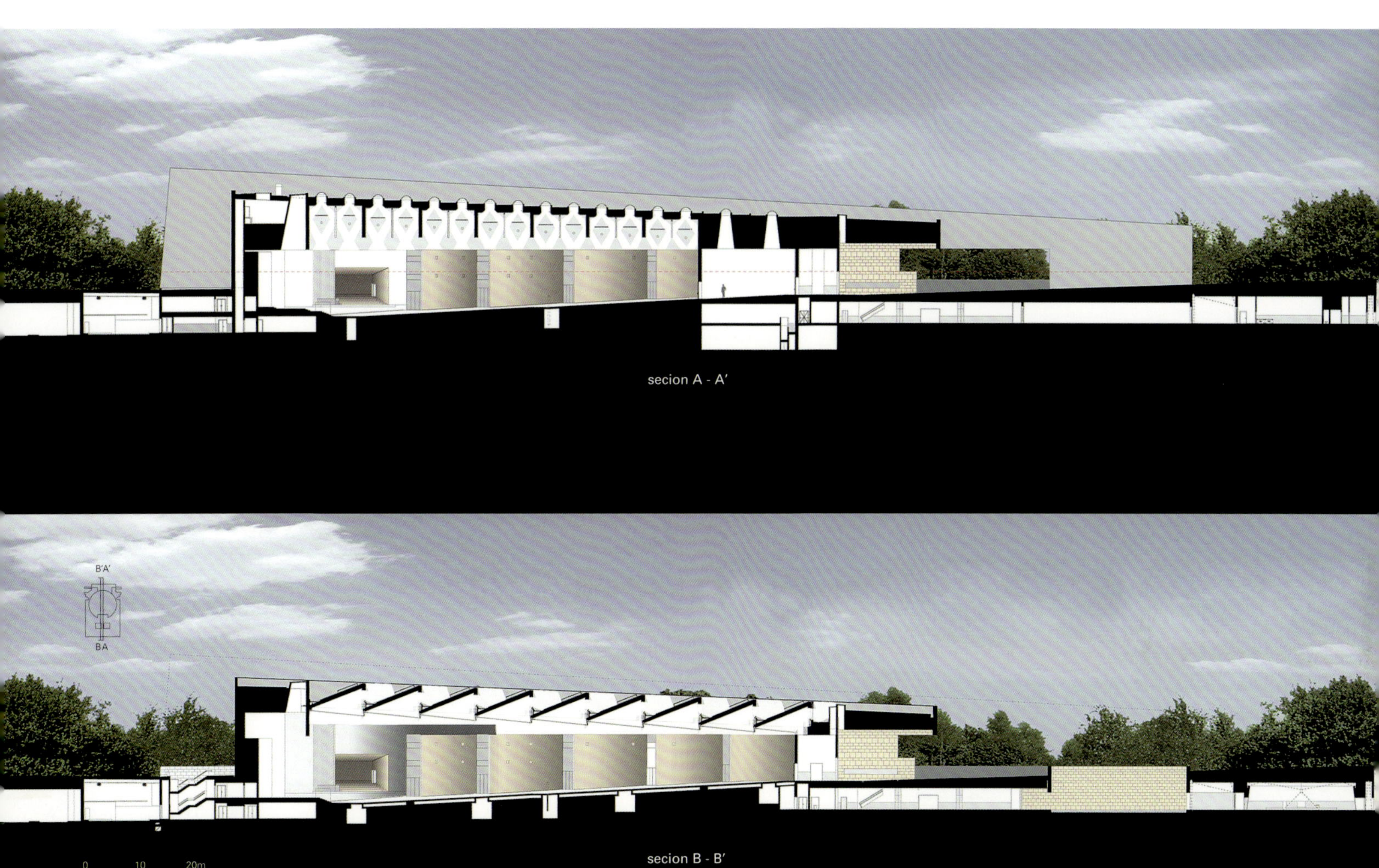
secion A - A'
B'A'
BA
0
10
20m
secion B - B'

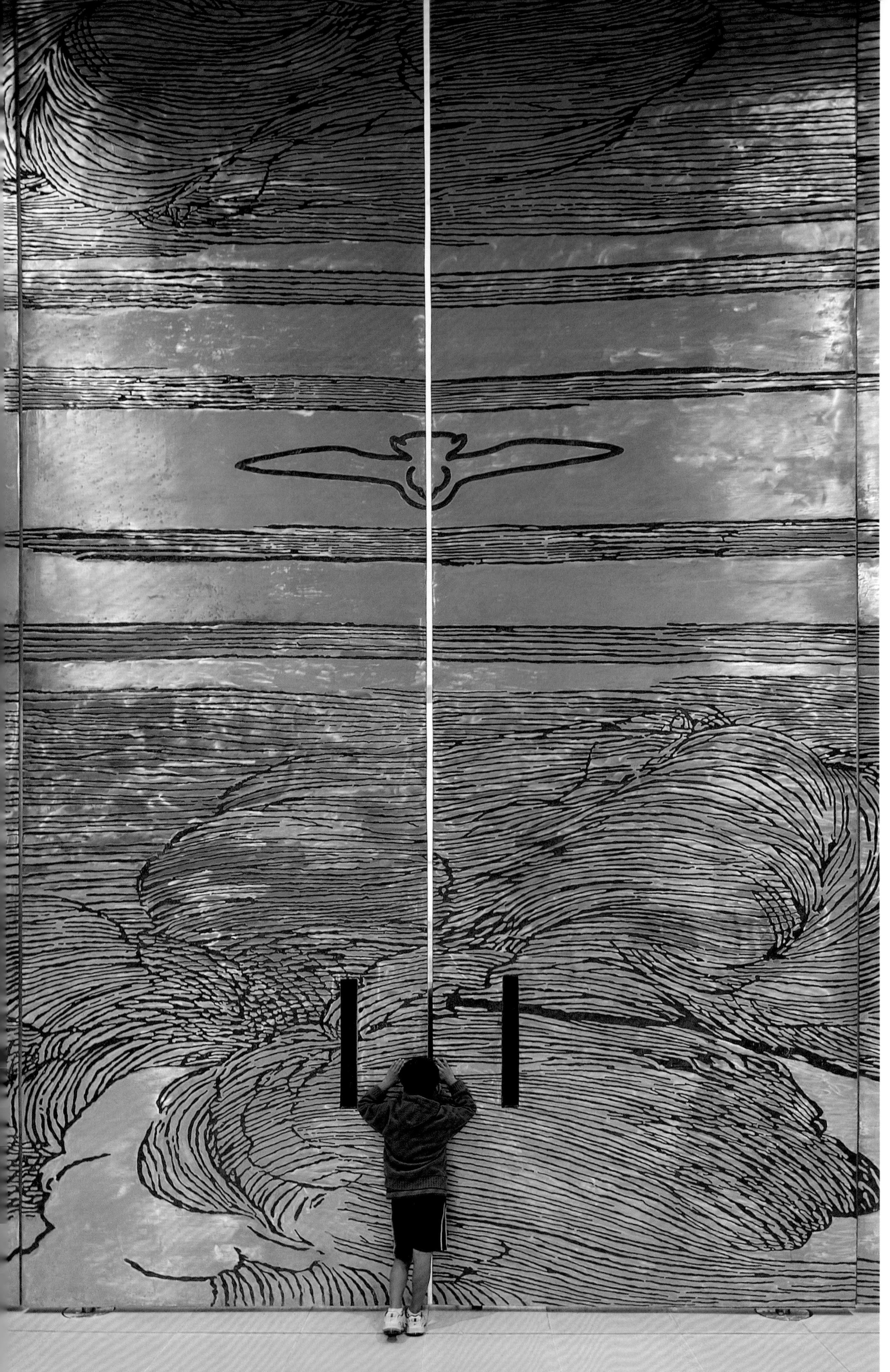

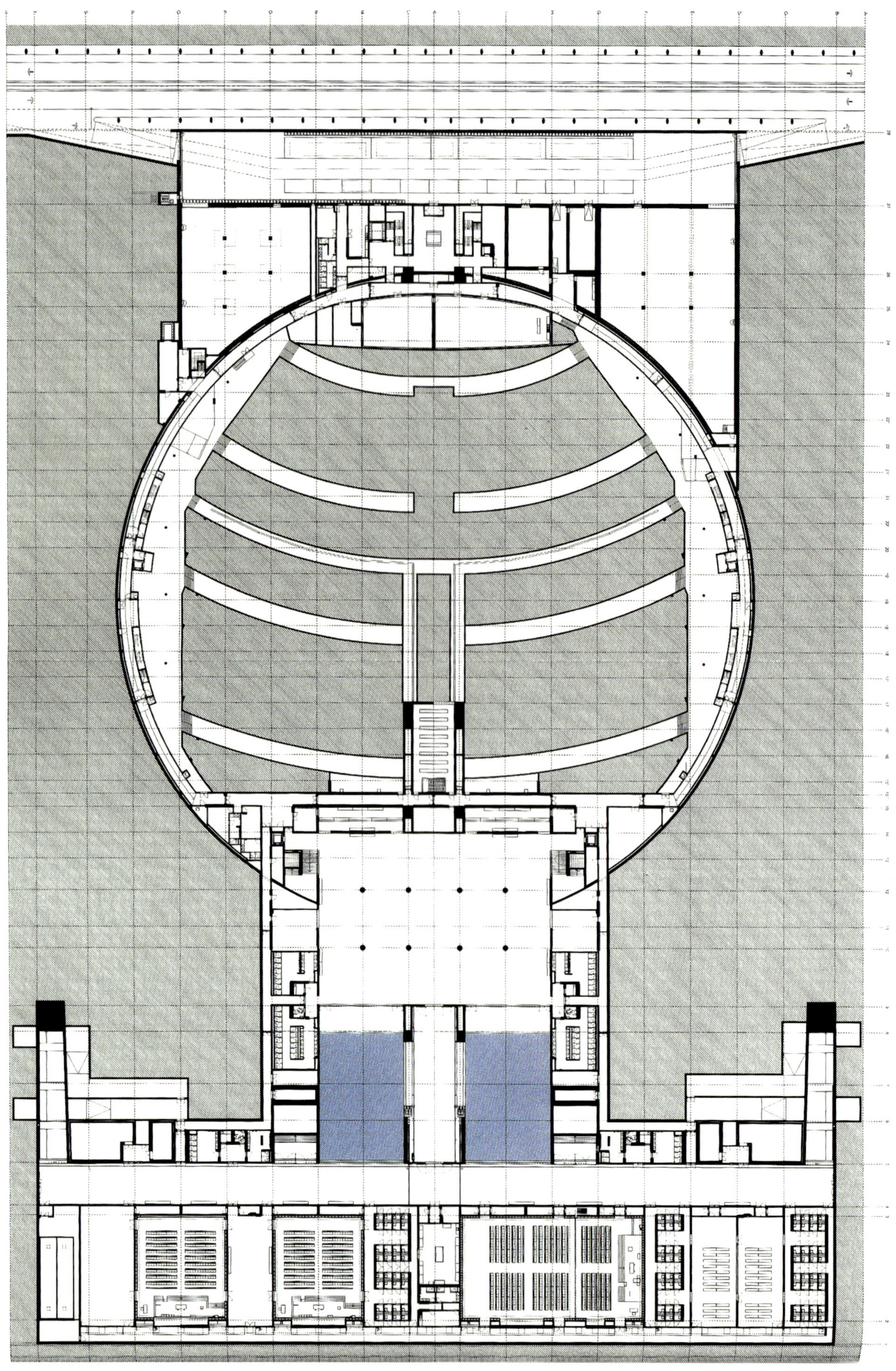
basement

0 5 10m

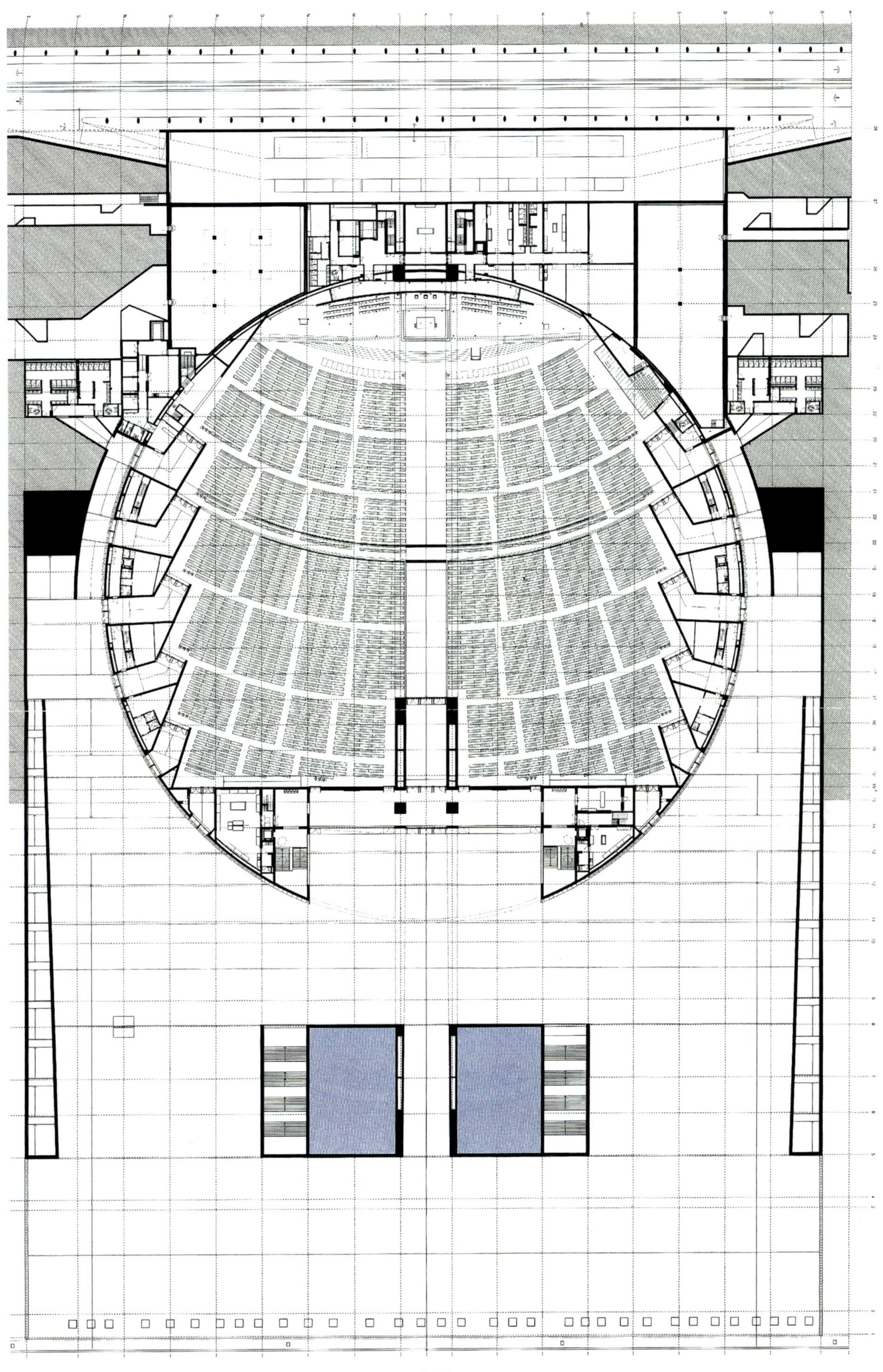
first floor

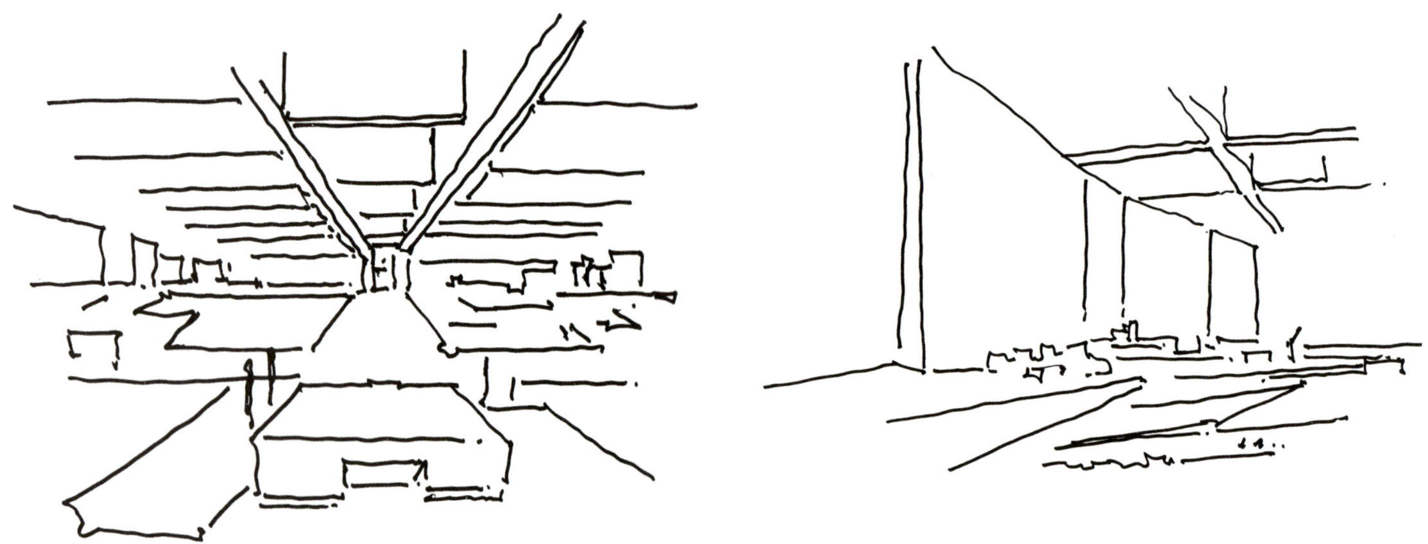

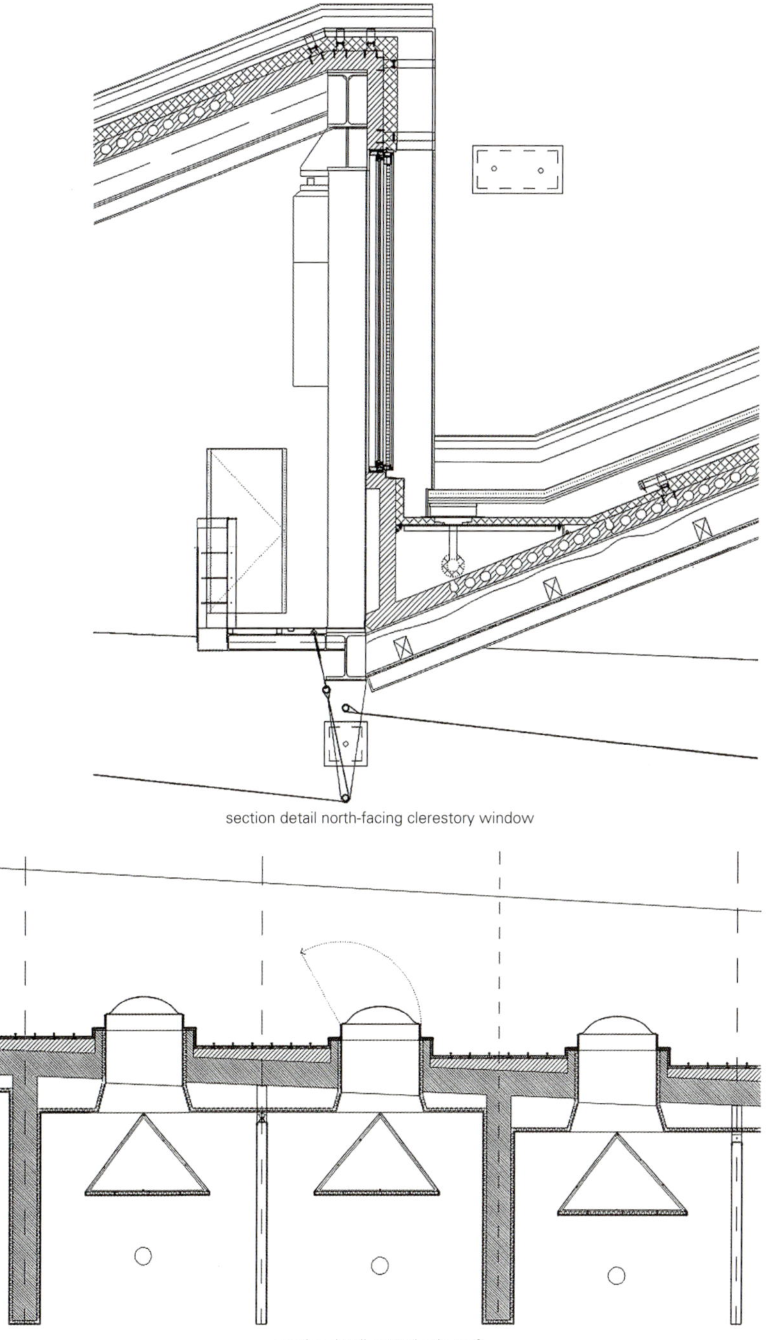

section detail north-facing clerestory window

section detail central axis roof

mural depicting by Alvaro Siza Viera

Bangladesh is a fertile alluvial land in the Gulf of Bengal and the land with the highest population density in the world and over 80% of the population live in rural areas. Much of the vernacular built tradition uses earth and bamboo as a building material, however, construction techniques are error-prone and many buildings lack foundations and damp proof coursing. Such buildings require regular maintenance, are often prone to damage and last on average only 10 years. It is particularly important to improve the quality of living in the rural areas in order to counteract the continuing population migration to the cities. The primary potential for developing building in the rural areas is the low cost of labour and locally available resources such as earth and bamboo. The project's main strategy is to communicate and develop knowledge and skills within the local population so that they can make the best possible use of their available resources.

METI Modern Education and Training Institute aims to promote individual abilities and interests taking into account the different learning speeds of the schoolchildren and trainees in a free and open form of learning. It offers an alternative to the typical frontal approach to lessons. The architecture of the new school reflects this principle and provides different kinds of spaces and uses to support this approach to teaching and learning.

On the ground floor with its thick earth walls, three classrooms are located each with their own access opening to an organically shaped system of 'caves' to the rear of the classroom. The soft interiors of theses spaces are for touching, for nestling up against, for retreating into for exploration or concentration, on one's own or in a group. The upper floor is by contrast light and open, the openings in its bamboo walls offering sweeping views across the surroundings, its large interior providing space for movement. Light and shadows from the bamboo strips play across the earth floor and contrast with the colourful materials of the saris on the ceiling.

The building rests on a 50 cm deep brick masonry foundation rendered with a facing cement

방글라데시의 손으로 만든 학교

방글라데시는 벵골 만을 끼고 있는 비옥한 충적토를 자랑하는 세계에서 인구밀도가 가장 높은 나라로 인구의 80% 이상이 농촌에 거주한다. 이 지역의 전통적인 건축은 대부분 흙과 대나무를 사용하여 짓는다. 그러나 시공방법에 문제가 많고 건물에 기초와 방습 장치가 없어 정기적인 보수가 필요하며, 쉽게 훼손되고 평균 수명이 10년에 불과하다.

도시로의 지속적인 인구 이동을 막기 위해서는 농촌 생활의 질을 개선하는 것이 특히 중요하다. 농촌 주택 개발의 일차적인 잠재력은 흙, 대나무와 같은 풍부한 자원과 낮은 인건비에 있다. 이 프로젝트의 주요 전략은 방글라데시 농촌의 자원을 최대한으로 활용할 수 있는 지식과 기술을 개발하여, 이 지역의 건설관계자들에게 전달하는 것이다.

METI Modern Education and Training Institute는 자유롭고 개방적인 학습 환경에서 학생들 개개인의 학습속도를 고려하여 능력과 흥미를 개발하는데 목표를 두고 전형적인 학습방법을 탈피한 하나의 대안을 제시하고자 한다. 새로운 학교는 이와 같은 교육 및 학습 방법을 지원하기 위해 다용도의 공간으로 설계되었다.

두꺼운 흙벽을 둘러친 1층에는 세 개의 교실이 있으며, 각 교실 뒤쪽은 유기적인 형태의 '동굴'로 이어진다. 교실 내부는 온화하며 촉감이 좋고 서로 부대끼며 놀 수 있을 만큼 아늑하다. 또한 개인 혹은 여럿이 탐구학습이나 집중학습을 할 수 있는 조용한 분위기를 연출한다. 반면 2층 공간은 밝고 개방적이며, 대나무 벽의 창문을 통해 탁 트인 주위 경관을 볼 수

Anna Heringer + Eike Roswag

School Handmade in Bangladesh

plaster. Bricks are the most common product of Bangladesh's building manufacturing industry. Bangladesh has almost no natural reserves of stone and as an alternative the clayey alluvial sand is fired in open circular kilns into bricks. These are used for building or are broken down for use as an aggregate for concrete or as ballast chippings. Aside from the foundation, the damp proof course was the other most fundamental addition to local earthen building skills. The damp proof course is a double layer of locally available PE-film. The ground floor is realised as load-bearing walls using a technique similar to cob walling. heaped on top of the foundation wall to a height of 65 cm per layer. Excess material extending beyond the width of the wall is trimmed off using sharp spades after a few days. After a drying period of about a week the next layer of cob can be applied. In the third and fourth layers the door and window lintels and jambs were integrated as well as a ring beam made of thick bamboo canes as a wall plate for the ceiling.

The ceiling of the ground floor is a triple layer of bamboo canes with the central layer arranged perpendicular to the layers above and beneath to provide lateral stabilization and a connection between the supporting beams. A layer of planking made of split bamboo canes was laid on the central layer and filled with the earthen mixture analogue to the technique often used in the ceilings of European timber-frame constructions. The upper storey is a frame construction of four-layer bamboo beams and vertical and diagonal members arranged at right angles to the building. The end of the frames at the short ends of the building and the stair also serve to stiffen the building. These are connected via additional structural members with the upper and lower sides of the main beams and equipped with additional windbracing on the upper surface of the frame. A series of bamboo rafters at half the interval of the frame construction beneath provide support for the corrugated iron roof construction and are covered with timber panelling and adjusted in height to provide sufficient run-off.

있다. 널찍한 내부는 운동 공간으로 사용할 수 있다. 대나무 조각에 의해 연출되는 빛과 그림자는 흙으로 만든 바닥을 수놓으며 천장의 화려한 전통 의상 옷감과 대조된다. 건물은 시멘트 플라스터로 만든 높이 50cm의 벽돌 기초에 놓여있다. 벽돌은 방글라데시의 건축산업에서 가장 일반적인 재료이다. 방글라데시는 천연 석재 자원이 거의 없기 때문에, 그 대안으로 점토질의 모래를 뚜껑이 없는 둥근 가마에 구워 벽돌을 만든다. 이 벽돌은 그대로 건물에 쓰거나 아니면 분쇄하여 콘크리트 골재용이나 바닥재로 사용한다. 벽돌 기초와 함께 이 프로젝트의 또 다른 중요한 기술은 방습처리이다. 방습 장치는 현지에서 구할 수 있는 폴리에틸렌 수지로 된 이중 켜로 해결하였다. 1층 벽은 볏짚을 섞은 흙을 물에 개서 내력벽을 쌓았다. 이 재료를 기초벽 위에 65cm의 높이로 쌓아올리고, 약 1주일 동안 마르도록 놔둔 후 그 다음 켜를 쌓는다. 세 번째와 네 번째 켜에는 문과 창호의 가로대와 세로대, 두꺼운 대나무로 만든 원형 대를 천장 판과 통합하였다.

1층 천장은 3중 대나무 켜로 되어 있으며 가운데 켜를 상하 켜와 수직으로 배열하여 측면을 고정하고 지지보들을 연결하였다. 그리고 유럽 목구조식 천장을 응용하여 쪼갠 대나무로 만든 판자를 가운데 켜에 두고, 흙으로 그 틈새를 메웠다. 2층 구조는 4중의 대나무 보와 건물에 직각으로 배열되는 수직 및 대각선 부재로 구성된다. 이 구조는 건물의 짧은 쪽 입면의 양 끝에서 끝나며 계단과 함께 건물을 강화한다. 대나무 서까래는 1층 구조 간격의 절반 길이로 주름 강판 지붕을 지지한다. 이 서까래는 판자로 덮이며 효과적인 배수를 위해 높이를 적절히 조정하였다.

Structural engineer Ziegert Roswag Seiler Architekten Ingenieure Burogemeinschaft - Christof Ziegert, Uwe Seiler **Bamboo construction** Emmanuel Heringer (basket weaver and carpenter), Stefanie Haider (blacksmith) **Constructor** Dipshikha/METI with local labourers and own training workshop **Client** Dipshikha/METI, Bangladesh in cooperation with Partnerschaft Shanti - Bangladesch e.V. and the Papal Children's Mission (PMK) **Location** Bangladesh **Bldg. area** 275 m^2 **Gross floor area** 325 m^2 **Construction** 2005 **Photograph** Courtesy of the architect ©Kurt Hörbst - p.280~281, 282 bottom, 283, 284, 285, 289, 290, 292, 293, 294~295, building team - p.282 top-left, 282 top-right, 286, 288

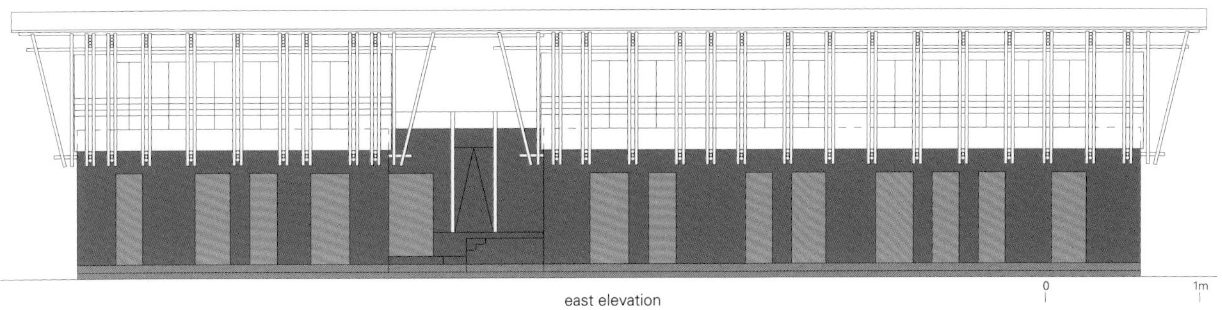

east elevation

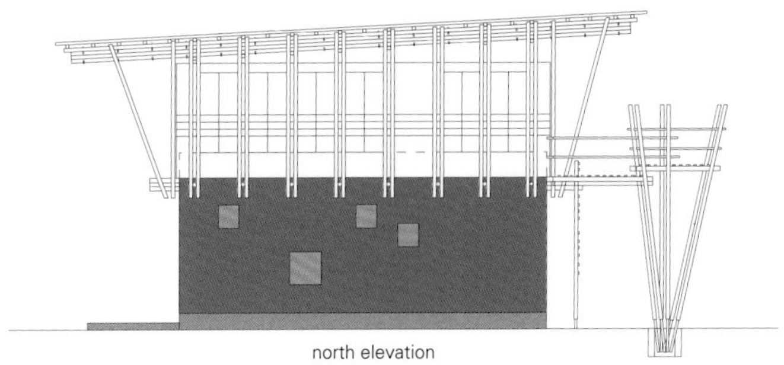
north elevation

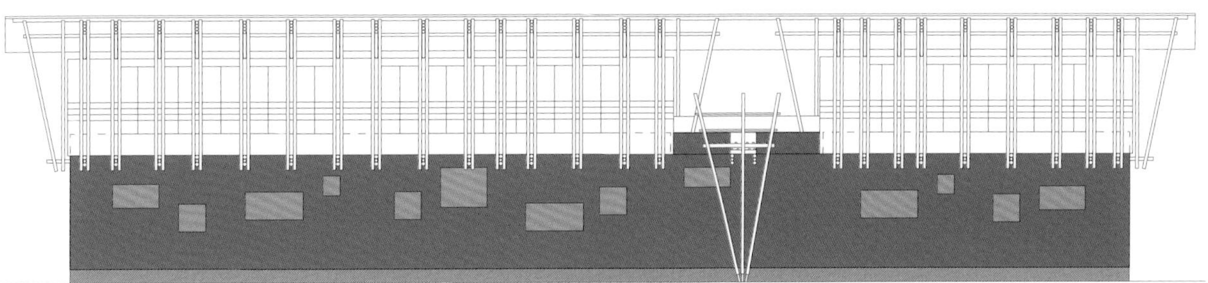
west elevation

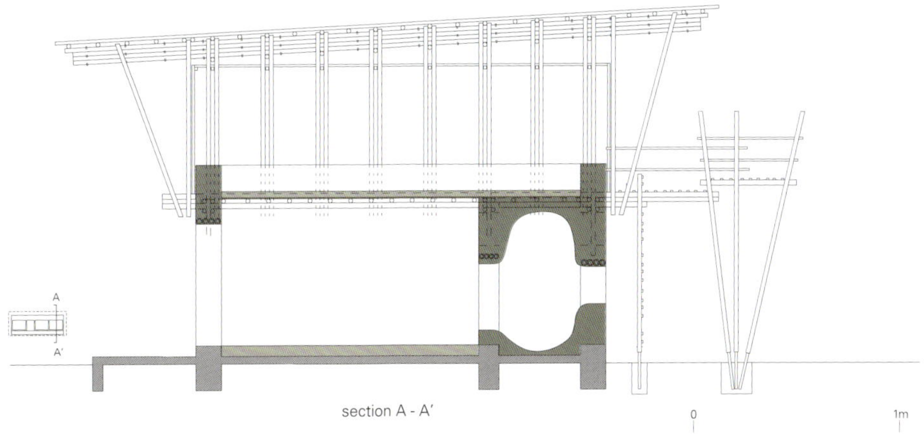

section A - A'

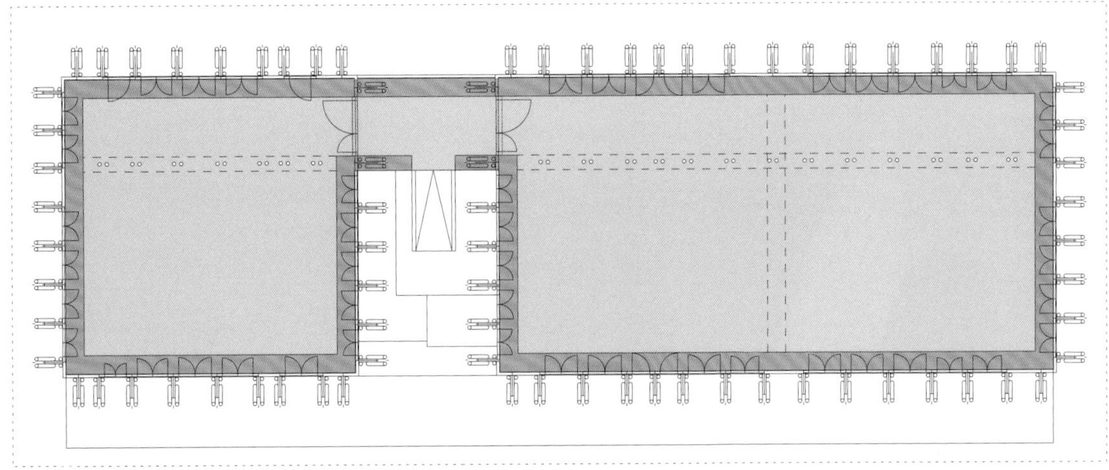
second floor

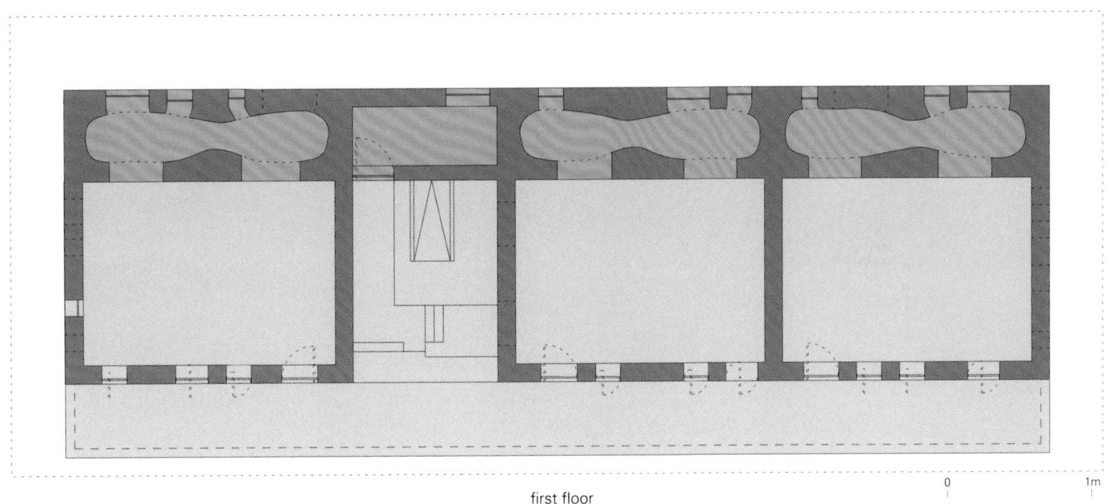

first floor

Alexandros N. Tombazis

©N. Kokkalias

Alexandros N. Tombazis was born in 1939 in India and graduated from the School of Architecture, National Technical University of Athens in 1962 with honours. He lives and works in Athens. Since 1963 he has been a principal in "Alexandros N. Tombazis and Associates Architects" (Meletitiki Ltd.). Participated on a number of occasions as a member of different juries of international and national architectural competitions, such as the E.C. DG XII Competitions, EUROPAN Competitions, PLEA Competitions etc. Also a member of different committees related to the use of alternative energy sources and bio-climatic architectural design. In 1991 he was elected an honorary fellow of the American Institute of Architects and in 2006 he was awarded an honorary phD by the Aristotle University of Thessaloniki. He has given lectures on many occasions worldwide and was a keynote speaker in UIA 2005 World Congress of Architecture.

알렉산드로스 톰바지스는 1939년 인도에서 태어나 1962년 아테네 국립기술대학 건축과를 졸업했으며, 현재 아테네를 중심으로 활동하고 있다. 1963년에 알렉산드로스 톰바지스 앤 어소시에이츠 아키텍츠를 설립했다. 다수의 국내외 건축 공모전에 심사위원으로 참여했으며 대체 에너지 자원의 사용과 생태기후적 건축 디자인 위원회의 회원이기도 하다. 1991년에는 미국건축가협회의 명예회원으로 선정되었으며 2006년에는 아테네의 아리스토텔레스 대학에서 명예박사학위를 받았다. 2005 UIA 세계건축회의의 기조연설을 비롯한 국제적인 많은 행사에서 강연을 해왔다.

Álvaro Leite Siza Vieira

Álvaro Leite Siza Vieira was born in Porto, Portugal in 1962. He graduated from the Faculty of Architecture in Escola do Porto in 1994. He collaborated with his father, architect, Alvaro Siza Vieira on a first project in Lisbon in 1987, a competition for the Museum of Contemporary Art in Helsinki in 1992 and the Olympic complex of the University of Palermo in 1995. He has participated in competitions, exhibited his own works at international design exhibitions, and designed many houses around Portugal. His recent projects include the Sports Complex in Porto (2004~) and the Museum of Manuel Cargaleiro Foundation (2003~).

알바로 레이테 시자 비에이라는 1962년 포르투갈 포르투에서 태어나 1994년 포르투 대학 건축학부를 졸업했다. 첫 번째 작업을 1987년 리스본에서 아버지인 알바로 시자와 함께 했다. 1992년에는 헬싱키 현대미술관 설계공모전에 참여했고, 1995년에는 빨레르무 대학의 올림픽 복합시설을 설계했다. 그 밖에 많은 설계공모전에 참가했으며, 국제적인 디자인 전시회에서 자신의 작업을 선보였다. 또한 포르투갈 지역에 많은 주택들을 설계했다. 최근작으로는 포르투의 스포츠 복합시설 2004~과 마누엘 까르갈레이루 재단 박물관 2003~ 등이 있다.

Anna Heringer + Eike Roswag

©Kurt Hörbst ©Marco Limberg

Anna Heringer[left] was born in Rosenheim, Germany in 1977. Currently living in Salzburg and Linz, Austria. She studied architecture at Linz University of the Arts, Austria. Since 2004 she has held a lectureship there, and is project manager at BASE - habitat/architektur konzepte, Linz University of the Arts. In 2006 she began her doctoral studies at Munich Technical University, on strategies for sustainable building in northern Bangladesh. She is vicechairwoman of Shanti, a German-Bangladeshi partnership founded in 1983, with the aim of arranging exchange programmes, for example the transfer of professional volunteers. She was awarded from the Bavarian Academy for Rural Areas (Bayrischen Akademie für den ländlichen Raum), 2006.

Eike Roswag[right] was born in 1969, Gießen and currently living in Berlin. He completed his architectural studies at Berlin Technical University in 2000, after which he took on freelance architectural work and consultancies. In 2003 he joined ZRS Architects and Engineers to plan and build a variety of projects using earth as a building material. In 2006 he joined the staff of Berlin Technical University and founded Roswag & Jankowski Architects Partnership.

아나 헤링어[왼쪽]는 1977년 독일 로젠버그 생으로 현재는 오스트리아 린츠와 잘츠부르크에 살고 있다. 2005년부터 린츠 예술대학의 BASE-해비타트/아키텍투르 콘첩트의 프로젝트 매니저이다. 1997년 방글라데시 협회에서의 자원봉사를 시작하여 2004년부터는 샨티 파트너쉽의 부의장을 맡고 있다. 2006년에는 뮌헨 공대에서 박사학위를 취득했으며 바이에른 아카데미에서 수여하는 농촌지역 상을 받았다.

아이케 로즈바그[오른쪽]는 1969년 독일 기센 생으로 현재 베를린에 살고 있다. 베를린 공대에서 건축을 공부하고, 공학박사 학위를 취득했다. 2003년 ZRS 아키텍츠 & 엔지니어링에서 흙을 건축 재료로 사용하는 다양한 프로젝트를 수행했다. 2006년에는 베를린 공대에서 근무하였고, 로즈바그 & 얀코프스키 아키텍츠 파트너쉽을 설립하였다.

Carme Pinós

Carme Pinós graduated at the Escuela Técnica Superior de Arquitectura de Barcelona in 1979. From 1982 she formed a partnership with Enric Miralles, which ended in 1991. The work of Pinós-Miralles received awards on several occasions, counting the FAD prize for the La Llauna School and the Igualada Cemetery Park, as well as the City of Barcelona Prize for the Archery Range Buildings for the 1992 Olympic Games.

In 1991 she set up her own studio and transferred the supervision and construction of several projects initiated in her previous office. Since then she has combined her activity as an architect with teaching, working as a guest professor, amongst others, at the University of Illinois at Urbana-Champaign (1994-1995), the Columbia University in New York (1999), the ETSAB in Barcelona (2002), the Harvard University Graduate School of Design (2003).

Projects having a great impact on her career are the Pedestrian Bridge in Petrer (Alicante), the Juan Aparicio Waterfront in Torrevieja (Alicante), the Cube Tower in Guadalajara (Mexico), and the Primary School in Castelldefels (Barcelona).

Carme Pinós has received the 1st Prize of the Biennial of Spanish Architecture in 2007 for the Cube Tower.

까르메 삐노스는 1979년 바르셀로나 상급 건축학교[ETSAB]를 졸업했으며 엔릭 미라예스와 함께 1982년 스튜디오를 설립했다. 이 시기에 설계한 이괄라다 납골공원과 1992년 올림픽 양궁연습장, 라 라우나 학교 등으로 다수의 상을 받았다. 1991년 자신의 사무소를 설립하면서 일리노이 주립대[1994~1995], 콜럼비아 대학[1999], ETSAB[2002], 하버드 디자인 대학원[2003] 등에서 학생들을 가르쳤다. 주요 작으로는 알리칸테의 삐뜨레르 인도교, 후앙 아빠리씨오 워터프런트, 멕시코 과달라하라의 큐브 타워, 까스뗄데펠스의 초등학교 등이 있다. 큐브 타워로 2007년 스페인 건축 비엔날레에서 1등상을 수상한 바 있다.

Fernando Martin Menis

Fernando Martin Menis was born in 1951 in Santa Cruz de Tenerife. He studied architecture at ETSA Barcelona and worked at Ricardo Bofill Studio in 1976. Participated in the Competition 'Parc de la Villette - Park and Center of Technology and Sciences' in 1980. The following year, he founded a Artengo, Menis, Pastrana Architects. Since July 2004, Menis has been working independently with his own studio.

He is a visiting professor and leads workshops and lectures in Vienna, Graz, Havana, Berlin, Paris, Roma, Barcelona, Valencia, Leon, Castellon, Santiago de Compostela, Benicassim, Las Palmas de Gran Canaria, Santander, Adeje, Las Palmas de Gran Canaria. Magma Arte & Congresos was nominated for European Prize Mies van der Rohe 2007, the Spanish Architecture Award 2007 and was a finalist at The IX Spanish Biennale of Architecture and Urbanism.

페르난도 마틴 메니스는 1951년 산타크루즈 데 떼네리페 생이다. 바르셀로나 상급 건축학교[ETSAB]에서 건축을 공부했고 1976년 리까르도 보필 사무소에서 일했다. 1980년 라빌레뜨 공원과 과학기술센터 설계 공모전에 참여했고 이듬해 AMP 아키텍츠를 설립했다. 2004년 7월부터 개인 사무소를 열어 운영하고 있다.

초청 교수로 다수의 강연을 했으며 비엔나와 그라츠, 하바나, 베를린, 파리, 로마, 바르셀로나 등지에서 워크숍을 진행하고 있다. 마그마 예술센터 및 회의장으로 2007년 미스 반 데 로에 상과 2007 스페인 건축상의 후보로 지목되었으며 제9회 스페인 건축 비엔날레의 최종 후보로 선정되기도 했다.

Herzog & de Meuron

Jacques Herzog, Pierre de Meuron, Harry Gugger (from left)

In 1978 Jacques Herzog and Pierre de Meuron established their office and became Herzog & de Meuron in 1997. Harry Gugger and Christine Binswanger joined the practice as partners in 1991 and 1994 respectively, followed by Robert Hosl and Ascan Mergenthaler in 2004, Stefan Marbach in 2006, and most recently, Wolfgang Hardt, David Koch and Markus Widmer in 2008. Currently the practice employs 19 associates and more than 270 architects working on over 40 projects worldwide. Branch offices are in London, Hamburg, Barcelona, New York and Beijing. Herzog & de Meuron has been awarded numerous prizes including The Pritzker Architecture Prize (2001), RIBA Royal Gold Medal (2007) and the Praemium Imperiale (2007).

Jacques Herzog and Pierre de Meuron are both teaching at Harvard University (since 1994) and at the Swiss Federal Institute of Technology Zurich, ETH Studio Basel (since 1999). In 2002 they co-founded the ETH Studio Basel - Contemporary City Institute.

Most recognized projects include Prada Aoyama Epicenter in Tokyo (2003); the Forum 2004 in Barcelona (2004); and Allianz Arena in Munich (2005). The latest additions are the new designs for the Parrish Art Museum in Southampton (projected completion 2009) and for the Miami Art Museum, Florida (projected completion 2010).

헤르조그 & 드 뫼론은 쟈끄 헤르조그와 피에르 드 뫼론이 1978년에 설립하였다. 현재 전세계 40여 개 프로젝트를 270명이 넘는 건축가들이 작업하고 있다. 또한 런던, 함부르크, 바르셀로나, 뉴욕, 베이징에 지사를 두고 있다. 헤르조그 & 드 뫼론은 2001년 프리츠커 상과 2007년 영국 RIBA에서 수여하는 로얄 골드 메달, 일본 예술협회의 프리미엄 임페리얼 상을 비롯하여 다수의 상을 받았다. 쟈끄 헤르조그와 피에르 드 뫼론은 1994년부터 하버드 대학에서, 1999년부터는 취리히의 스위스 연방과학기술원과 바젤 ETH 스튜디오에서 학생들을 가르치고 있으며, 2002년 바젤 ETH 스튜디오에 현대도시연구소를 공동 설립했다.

주요작으로는 동경의 프라다 아오야마 에피센터, 바르셀로나 2004 포럼, 뮌헨 알리안츠 경기장이 있다. 현재 사우스햄튼의 패리쉬 예술박물관과 플로리다의 마이애미 예술박물관의 설계를 진행하고 있다.

Miralles Tagliabue EMBT

In 1998 Benedetta Tagliabue and Enric Miralles founded Miralles Tagliabue EMBT in Barcelona. Their work together includes very representative buildings and public spaces in Barcelona as well as in other European cities. EMBT's architecture has been awarded multiple international awards, Premio Nacional de Catalunya, RIBA Best Building Award, Premio FAD de Arquitectura, Rietveld Prize, BDA Architecture Preis between them.

Their recent completed projects include Gas Natural Headquarters (2007), Office Building in Barcelona, Public Library in Palafolls and Hafencity Hamburg Public Space (2006), Rectorship Building for the Vigo University, Spain (2005), Santa Caterina Market in Barcelona, Spain, Scottish Parliament in Edinburgh, Scotland (2004).

미라예스 따글리아부 EMBT는 1998년에 설립되었으며, 바르셀로나를 비롯하여 유럽 여러 도시의 공공 장소에 있는 중요 건축물들을 작업해왔다.

EMBT의 건축은 까딸루냐 국제건축상, RIBA 최고건축상, FAD 건축상, 리트벨트 상, BDA 건축상 등 국제적인 건축상을 수상한 바 있다.

최근작으로는 바르셀로나의 천연가스 본사, 팔라폴스 공공 도서관, 하펜시티 공공 광장, 비고 대학교 본관, 까떼리나 시장 재건축, 스코틀랜드 의회 등이 있다.

Morphosis

©Mark Hanauer

Morphosis was founded in 1972 in Los Angeles as an interdisciplinary and collevtive practice involved in experimental design and regorous research.

Named after the Greek term, morphosis, meaning to form or be in formation, Morphosis is a dynamic and evolving practice that responds to the shifting and advancing social, cultural, political and technological conditions of modern life.

Thom Mayne is the founder and principal of Morphosis. He was born in Connecticut in 1944 and received his Bachelor of Architecture degree from the University of Southern California in 1968. In 1978 he received his Master of Architecture from Harvard University. He has held teaching positions at Columbia University, Harvard University (Elliot Noyes Chair, 1998), Yale University (Eliel Saarinen Chair, 1991), the Berlage Institute in the Netherlands and the Bartlett School of Architecture in London. His distinguished honors include Cooper-Hewitt, National Design Award (2006), Pritzker Prize Laureate (2005), the 2000 American Institute of Architects / Los Angeles Gold Medal in Architecture, and the Chrysler Design Award of Excellence (2001).

1972년 로스앤젤레스에 설립된 모포시스는 여러 분야에 걸쳐 실험적인 디자인과 정밀한 연구를 수행하는 사무소이다. 그리스어로 '형성하다'라는 의미인 모포시스는 다변하는 현대의 사회, 문화, 정치, 기술적 조건에 역동적이고 진보적으로 반응하는 회사이다.

모포시스의 설립자이자 대표인 톰 메인은 1944년 코네티컷에서 태어났다. 1968년 남 캘리포니아 대학을 졸업하고 1978년에는 하버드 대학에서 박사학위를 받았다. 콜럼비아 대학과 하버드 대학, 예일 대학, 네덜란드의 베를라헤 건축대학과 영국 바틀렛 건축학교에서 학생들을 가르쳤다. 2006년 쿠퍼 휴잇 국립 디자인 상, 2005년 프리츠커 상, 2000년 미국건축가협회 로스앤젤레스 지역 골드메달과 2001년에는 크라이슬러 디자인 우수상을 수상한 바 있다.

Paulo David

Paulo David was born in Funchal, Portugal in 1959. He graduated from Faculdade de Arquitectura da Universidade Tecnica de Lisboa with Diploma in Architecture in 1989. He established an Office of Paulo David Arquitectos in Funchal, 1996. He was a visiting professor in University of Arte and Design, Madeira from 2001 to 2004. His major projects include Eduardo Jesus House in 2003, Arts Centre Casa das Mudas, Salinas Swimming Pools and Volcanism Centre Sao Vicente in 2004. Arts Centre Casa das Mudas in Calheta, Madeira was selected for Mies Van der Rohe Award 2005.

파울로 데이비드는 1959년 포르투갈 훈샤우 생으로 리스본 건축대학을 졸업했다. 1996년 훈샤우에 자신의 사무소인 파울로 데이비드 아키텍토스를 설립했다. 2001년부터 2004년까지 마데이라 예술 디자인 대학에서 초청교수로 있었다. 주요 작품으로는 2003년의 에두아르두 제수스 주택과 까사 다스 무다스 예술센터, 살리나스 수영장과 상 비센뜨 화산센터 등이 있다. 까사 다스 무다스 아트센터로 2005년 미스 반 데 로에상을 수상했다.

Peter Zumthor

©Ebner Gerry

Peter Zumthor was born in Basel in 1943, trained as a cabinetmaker, designer, and architect at the Kunstgewerbeschule Basel and Pratt Institute, New York. In 1979 he opened his own practice Atelier Zumthor in Haldenstein, Switzerland. He is a professor at the Accademia di architettura, Universita della Svizzera italiana.

His major projects include Swiss Pavilion Expo 2000, Hannover, Documentation Center "Topography of Terror," Berlin, constructed parts of 1997 demolished 2004 by Berlin State, Art Museum Kolumba, Cologne, 2007 and Saint Bruder Klaus Field Chapel, Scheidtweiler farm, Mechernich, Germany, 2007.

피터 줌터는 1943년 스위스 바젤에서 태어나 바젤 미술공예학교와 뉴욕의 프랫 인스티튜트에서 가구제작, 디자인, 건축을 공부했다. 1979년 스위스 할덴슈타인에 자신의 사무소인 아뜰리에 줌터를 열었다. 현재 스위쩨라 이탈리아나 대학교 건축학부에서 학생들을 가르치고 있다.

주요작으로는 하노버 엑스포의 스위스관과 베를린 문헌정보센터 '공포의 지형도', 콜룸바 미술관과 성 클라우스 예배당 등이 있다.

Toyo Ito

Toyo Ito graduated from The University of Tokyo, Department of Architecture in 1965. He started his own studio, Urban Robot (URBOT) in Tokyo, 1971 and changed its name to Toyo Ito & Associates, Architects in 1979. Toyo Ito has been holding a professorship at the Tokyo Women's University. He is also an honorary professor at the University of North London and has served as guest professor at Columbia University. He is teaching at Tama Art University as a Visiting Professor. In 1998, Ito served as vice commissioner for Kumamoto Artpolis.
His highest profile projects include Serpentine Gallery Pavilion 2002, TOD'S Omotesando Building in Tokyo, Japan, 2004 and Tama Art University Library (Hachioji campus) in Tokyo and Japan, 2007.

이토 도요는 동경 대학 건축학부를 1965년 졸업했다. 1971년 자신의 스튜디오인 어반 로봇 URBOT을 개소하였고 79년에 도요 이토 앤 어소시에이츠 아키텍츠로 이름을 바꾸었다. 그는 동경 여자대학교의 교수이자 타마 예술대학의 초청 교수이다. 또한 북런던 대학에서 명예교수로 있었고, 콜럼비아 대학의 객원 교수이다. 1988년에는 구마모토 아트폴리스의 커미셔너를 맡았다. 주요 작품으로는 2002년 서펜타인 갤러리 파빌리온, 2004년 일본 동경의 토드 오모테산도 빌딩과 2007년 타마 예술대학 도서관이 있다.

3deluxe

The collective title 3deluxe connotes an interdisciplinary team of about 30 individuals from the fields of architecture, interior design, art, graphic design, media design and product design. On the basis of this broad spectrum of specialist knowledge, 3deluxe devises holistic design solutions that range from graphic identities via media installations to architecture and all boast a coherent aesthetics.
3deluxe's characteristic, organically inspired visual idiom has proven as much of a style icon in international context as the virtually expanded space concepts and the striking graphic design that redefines the interface between two- and three-dimensionality.
3deluxe was formed in 1992 in Wiesbaden by communication designers Andreas and Stephan Lauhoff as well as interior designer Nikolaus Schweiger and designer Dieter Brell. Since 1999 system modern gmbh - with industrial engineer Peter Seipp as MD - has been responsible for the development, production and management of 3deluxe's common projects.

©Isa Schafer

Stephan Lauhoff, Dieter Brell, Peter Seipp, Andreas Lauhoff (from left)

3디럭스는 약 30명의 건축, 인테리어 디자인, 예술, 그래픽 디자인, 미디어 디자인 및 제품디자인 분야의 전문가들로 구성되어 있다. 이러한 광범위한 전문 인력을 기반으로 그래픽과 미디어 설치, 건축과 모든 관련 분야에 이르는 전체적인 디자인 솔루션을 제안한다. 유기적인 비주얼 표현이 특징인 3디럭스는 2차원과 3차원을 넘나들며 확장된 공간 컨셉과 인상적인 그래픽 디자인과 같은 수많은 스타일 아이콘들을 선보여왔다.

3디럭스는 커뮤니케이션 디자이너 안드레아와 슈테판 라우호프 형제가 인테리어 디자이너인 니콜라우스 슈바이거, 디자이너 디에터 브렐과 함께 1992년 비스바덴에 설립했다. 1999년부터 산업디자이너인 피터 사이프가 MD로 참여한 시스템 모던 사가 3디럭스 프로젝트의 개발과 생산, 경영을 맡고 있다.